P9-DWP-251

The Math Instinct

The Math Instinct

Why You're a Mathematical Genius (Along with Lobsters, Birds, Cats, and Dogs)

by

Keith Devlin

Thunder's Mouth Press
New York

THE MATH INSTINCT
WHY YOU'RE A MATHEMATICAL GENIUS
(ALONG WITH LOBSTERS, BIRDS, CATS, AND DOGS)

Published by
Thunder's Mouth Press
An Imprint of Avalon Publishing Group Inc.
245 West 17th St., 11th Floor
New York, NY 10011

ISBN 1-56025-672-9

Book design by Maria Elias
Printed in the United States

Contents

Acknowledgments

Much of the account of vision in chapter 8 is based on the excellent summary of the human visual system given by Steven Pinker in his book *How the Mind Works.* Steve kindly gave me permission to use some of the illustrations from his book, and provided me copies of the original Postscript files. Those illustrations were all drawn by Ilavenil Subbiah. For anyone who wants to learn more about vision, I highly recommend Dr. Pinker's treatment of that topic.

Figures 1.1, 2.1, 4.1, 4.2, 4.3, 5.3, 5.4, 6.1, 6.3, 6.4, 6.6, 7.1, 7.2, 7.3, 7.4, 8.1, and 8.8 [f] were illustrated specifically for this book by Simon Sullivan.

Figure 5.3 originally accompanied the article "Measuring Beelines to Food," published in *Science,* Vol. 287, Issue 5454, 817-818, 4 February 2000, and is reproduced here by courtesy of the article's author, Professor Thomas Collett of the Sussex Centre for Neuroscience at the University of Sussex in England.

Figure 6.2 originally appeared as Figure 3.6 in *Mathematical Biology II, Spatial Models and Biomedical Applications* (Springer, New York 2003) by James D. Murray, now Emeritus Professor of Applied Mathematics at the University of Washington, Seattle, who graciously gave his permission to reuse the image here.

Figures 7.1, 7.2, 7.3, and 7.4 accompanied the article "How Animals Move: An Integrative View," published in *Science,* Vol 288, Issue 5463, 100-106, 7 April 2000, and are used here courtesy of the article's author, Professor Michael Dickinson of the Department of Bioengineering at Caltech, Pasadena.

Figures 11.1 and 11.2 appear by courtesy of Professor Denise Schmandt-Besserat of the College of Fine Arts and the Center for Middle Eastern Studies at the University of Texas at Austin.

• • •

I would like to thank my agent, Ted Weinstein, for his enthusiasm for this project and the many efforts he put into helping me shape the book into its present form and finding the right publisher. Thanks also to John Oakes of Thunder's Mouth Press, who is that right publisher. John wanted this book from the moment he first learned of it, and did everything he could to see it into print.

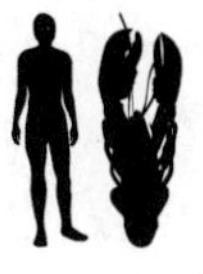

1.

OUT OF THE MINDS OF BABIES

In 1992, a young American researcher named Karen Wynn made an announcement that stunned child psychologists all around the world. Wynn claimed to have shown that babies as young as four months old could do simple addition and subtraction problems. In fact, other experimenters subsequently showed that babies can do the same math when they are only two days old!

How had Wynn done it? After all, four-month-old babies cannot yet talk, so how can we possibly discover whether they know that 1 + 1 = 2, to pick one of the examples Wynn claimed her young subjects could do? And how did Wynn manage to pose such a question in the first place so that the children could understand what she was asking?

Before I tell you how Wynn got around these problems, I should make it clear exactly what Wynn claimed to have discovered. First, she did not claim that her subjects had any conscious concept of numbers. As any parent knows, the counting numbers, 1, 2, 3, and so on, have to be taught to young children, and before that can happen the children have to learn how to use language, something that does not happen with a four-month-old baby. Rather, what Wynn claimed was that:

> 1. The children she examined could tell the difference between a single object, a pair of objects, and a collection of more than two objects.
> 2. They knew that if you take, say, two single objects and put them together, the resulting collection has exactly two objects in it, not one object and not three.
> 3. They knew that if you take, say, two objects and remove one of them, you are left with exactly one object; you don't end up with two objects or with none.

The normal way for an adult to describe these abilities would be to say that:

1. The children she examined knew the difference between the numbers 1 and 2 and the difference between 2 and any larger number.
2. They knew that, say, 1 + 1 = 2, and that 1 + 1 is not equal to 1 or 3.
3. They knew that, say, 2 – 1 = 1, and that 2 – 1 is not equal to 0 or 2.

Clearly, to express the ability this way requires an understanding of numbers, at the very least the numbers 0, 1, 2, and 3. Now, all the evidence that we have about the way the human brain handles numbers indicates that our ability to handle *numbers* only comes after the individual learns the numbers *words* "one," "two," "three," and so on. (Work with chimpanzees and other primates suggests that learning the number symbols "1," "2," "3," work just as well in this regard. The point is that acquisition of the *number concept* seems to require first having a word or symbol to refer to that concept.)

Strictly speaking, then, Wynn's claim was really about *numerosity,* by which I mean *number sense,* and especially a sense of the size of a collection, rather than numbers. What she was saying was that very young children have a reliable sense of the size of small collections of objects. But that did not lessen the surprise caused by Wynn's announcement. After all, everyone knew that four-month-old babies don't know how to use number words. Most experts assumed that a sense of numerosity developed

after the child learned how to count. Wynn was claiming that the number sense comes first. That meant that either we are born with such a sense, or at least we acquire it automatically within at most a few weeks of birth. (As we'll see below, subsequent research showed that if we are not actually born with a number sense, we acquire it within at most a few *days* of birth.)

Here is what Wynn did to arrive at her discovery. (Incidentally, Wynn's experiment has been repeated successfully many times over by different psychologists around the world, so there is no longer any doubt about the accuracy of her findings.)

The trick was to make use of the fact that even very young babies have acquired a fairly well developed sense of "the way things are." If a baby sees something that runs counter to its expectations, it will pay attention to it as it attempts to understand what it sees. By filming the child, particularly its eyes, as it is presented with various scenes, and then measuring the time the baby spends attending to each scene, the investigator can determine what runs counter to the baby's expectations. For example, if a baby is shown a series of pieces of fruit on plates, and is then shown an apple suspended in midair with no apparent means of support, the baby will stare measurably longer at the suspended fruit than it does at the fruit on plates.

Wynn sat her young subjects in front of a small puppet theater and set the (hidden) film camera rolling. (See

figure 1.1) The puppet stage was initially empty. The experimenter's hand came out from one side and placed a puppet on the stage. Then a screen came up, hiding the puppet. The experimenter's hand appeared again holding a second puppet, which it put behind the screen. Then the screen was lowered to reveal the two puppets. The child watched attentively throughout.

Figure 1.1. In the famous experiment carried out by psychologist Karen Wynn in 1992, a small child is shown correct and incorrect arithmetical sums played out on the stage of a puppet theater. By measuring the child's responses, as indicated by facial expressions, the experimenter can test whether the subject knows the difference between correct and incorrect arithmetic.

Wynn repeated the procedure several times in succession. On some repetitions, however, when the screen was lowered there was only one puppet on the stage. On other occasions, lowering the screen revealed three puppets.

(The experimenter had simply adjusted the stage out of the baby's sight.) Whenever the lowering of the screen revealed one puppet, or three, the baby stared longer than when there were the expected two. Having seen two puppets placed on an initially empty stage, one after the other, the baby clearly expected to see two puppets. When the outcome ran counter to this expectation, the baby was puzzled. On average, the child stared a full second longer when presented with an incorrect outcome than it did when presented with the correct outcome. The experiment showed that the baby "knew" that $1 + 1 = 2$ and that the additions $1 + 1 = 1$ and $1 + 1 = 3$ are both false. Similar experiments showed that the baby knew that $1 + 2 = 3$ as well.

Wynn obtained similar results when she modified the procedure to test the baby's understanding of subtraction. For example, the baby would initially be presented with two puppets on the stage. The screen would come up to hide the puppets, and the experimenter's hand would then appear and remove one puppet. The screen was then lowered to reveal either no puppets, one puppet, or two. The child stared longer at the stage—up to three seconds longer in some cases—when it saw two puppets or none than it did when there was exactly one puppet remaining. It "knew" that $2 - 1 = 1$ and that the subtractions $2 - 1 = 0$ and $2 - 1 = 2$ are both false. It also knew that $3 - 1 = 2$ and $3 - 2 = 1$.

Psychologists were stunned when Wynn announced her results, and many skeptical researchers around the world devised variants of her procedure to determine

whether her conclusions were correct. In particular, they wanted to see if Wynn was justified in concluding that the longer the subjects stared at arithmetically incorrect outcomes was really due to a sense of the size of a collection, numerosity, and not to some other cause.

One possibility was that it was not the number of objects that caused the different attention spans, but some feature of their physical arrangement. To test this particular possibility, Etienne Koechlin, a French psychologist, repeated Wynn's experiment, but with the puppets placed on a slowly revolving turntable. The constant movement of the puppets on the stage meant that the child could not form a fixed image of the scene, and could not predict the arrangement of objects it expected to see on the stage when the screen was lowered. Koechlin's results were exactly the same as Wynn's. The baby stared longer when presented with an arithmetically incorrect outcome than it did when the outcome was arithmetically correct. Koechlin's experiment eliminated any possibility that the child was responding to physical arrangement rather than quantity of units.

Another variation of Wynn's procedure was carried out by the American psychologist Tony Simon. In addition to confirming Wynn's original conclusion about numerosity, Simon uncovered another fascinating aspect of the way young children view their world.

When Simon performed the experiment, he sometimes changed the objects while they were behind the screen, exchanging, say, two red puppets for two blue

ones, or a red and a blue puppet for one or two yellow balls. The children showed no surprise when the screen was lowered to reveal that the objects had changed color or had changed from puppets to balls, *provided the arithmetic was correct.* Apparently, four-month-old babies are unfazed when they see objects change color or transform themselves into other objects, but they balk when they see two objects become one, or vice versa.

In other words, not only do very young children have a sense of number, but their expectation that number does not change seems to be more basic than their sense that color, shape, or appearance should not change. In another variant of Wynn's experiment, designed to test this view of the world, a baby was seated in front of a screen from behind which a red ball and a blue rattle alternately appeared. Provided the baby never saw the two objects together, it was quite happy to see just one of the two objects when the screen was lowered. It apparently accepted that objects can keep changing their appearance from moment to moment. This was true for babies up to one year of age. Only when a child is a year or more old does the successive appearance of two different-*looking* objects from behind the screen lead to an expectation that there are actually two different objects there.

I should make it clear that the number sense in infants that Wynn and subsequent experimenters observed was strictly limited to collections involving one, two, and three objects. For instance, children less than a

year old seemed unable to distinguish four objects from five objects. But as the various experiments showed, for collections of three objects or fewer, a four-month-old child has a well developed sense of numerosity and a basic understanding of addition and subtraction. When exactly does the child acquire them? Or is it born with them?

An experiment performed by the American psychologists Sue Ellen Antell and Daniel Keating showed that the ability to tell the difference between one object and a collection of two objects, and between two objects and a collection of three objects, is present in babies just a few days after birth. Antell and Keating adopted an experimental procedure first used by another American psychologist, Prentice Starkey. As with Karen Wynn's experiment, Starkey's procedure used the babies' visual attention span to see what they found surprising. The subjects were videotaped, so that the length of time they stared at a particular event could be accurately measured

In Antell and Keating's experiment, a baby just a few days old was shown slides projected onto a screen. The first slide showed two dots, side by side. When the slide first appeared, the baby stared at it for a while. Then it lost interest and its eyes started to wander. At that moment, the slide was replaced by one showing a slightly different arrangement of two dots. The child glanced back quickly, but soon lost interest. The slide was changed again to show two dots in still another arrangement. Again the child glanced back at the new arrangement but again it

rapidly lost interest. With each repetition of the procedure, the baby's rearoused gaze grew more and more brief. Then, suddenly, a slide appeared showing not two but three dots. Immediately the child's interest was aroused and it stared at the slide for a measurably longer period of time (jumping from 1.9 seconds to 2.5 seconds in one run of the experiment). Clearly, the subject had spotted the change in the number of spots. The same thing happened when the experiment began with the babies being shown three dots and the number was suddenly reduced to two.

By repeating the procedure many times, with dots arranged in different patterns, displayed in different orders, the experimenters eliminated any possibility that it was some change in appearance other than the change in the number of dots that was catching the babies' attention. Thus, the evidence was clear: Even a few days after they are born, babies have a sense of number.

Another experiment, conducted by the French psychologist Ranka Bijeljac, showed that newborn babies' sense of number is not restricted to collections that the baby sees. They can also tell the difference between two and three sounds heard in succession. Bijeljac used a different method of measuring the young subjects' attention span. Because the babies were being tested by sounds, it made little sense to videotape their faces and measure the length of their gaze. (There was nothing particular to gaze at.) Instead, Bijeljac made use of the babies' sucking reflex to monitor their interest. Each baby was provided

an artificial nipple to suck on. The nipple was connected to a pressure-sensing device that measured the amount the baby was sucking at each instant, sending the output to a computer. When the baby's interest was aroused, it sucked vigorously on the nipple. When its interest waned, it started to suck less.

The pressure sensor also controlled a device that generated prerecorded sounds, nonsense words of two or three syllables, such as "aki" or "bugaloo." A typical run of the experiment would proceed in this way: It did not take long for the baby to discover that when it sucked on the nipple, a sound was produced. Once it made that discovery, it started to suck vigorously, producing one sound after another. The apparatus was set up so that the nonsense words produced initially all had the same number of syllables, either two or three. After a while, the baby's interest waned, and its sucking slowed down. When the computer detected this drop, it switched over to producing nonsense words with a different number of syllables (from two to three, or vice versa). As soon as this happened, the baby started to suck vigorously again, producing more of the new-sounding words. Again, after a period of hearing the new kinds of words, the baby's interest dropped and its sucking pressure diminished, and once again the computer switched the number of syllables, arousing the child's interest once more. Since the change from one word to another did not arouse the baby's interest when the number of syllables remained the same, but only

occurred when there was a change in the number of syllables, the baby was seen to be responding to the *number* of syllables and not to some other feature of the sounds.

But there is more. Antell and Keating's research shows that, when each of us was just four days old, we were able to distinguish between collections of two and three *objects* that we saw. Bijeljac's results show that we could also distinguish between two and three *sounds* that we heard. Now, by the time we are adults, we have developed that early sense of numerosity to a more abstract level: We have an abstract sense of *twoness* and *threeness* that transcends any particular collection of things in the world. For instance, we recognize a similarity between a collection of two apples, two dots on a page, two elephants in a cage, two beats of a drum, and two airplanes in the sky. That twoness that all those collections have in common is a highly abstract sense of number. Indeed, our abstract sense of twoness, threeness, etc. is the beginning of mathematics. When did we acquire that deeper sense of number?

We certainly have the beginnings of a sense of twoness and threeness by the time we are six to eight months old. This was demonstrated by Prentice Starkey, the man who first designed the experiment used by Antell and Keating.

In an ingenious experiment, Starkey seated his subjects, babies aged between six and eight months, in front of two slide projectors arranged side by side. He videotaped

the children's faces to determine which projector interested them the most at any one time.

The two projectors simultaneously displayed pictures of a collection of two or three objects, arranged randomly. One projector would show a picture of two objects, the other a picture of three objects. Sometimes the projector on the left would show two objects and the one on the right would show three, and on other occasions the situation would be reversed.

As the two pictures were displayed, a loudspeaker situated between the two projectors played a sequence of two or three drumbeats. When the experiment began, the baby would pay attention to both pictures. Since the picture with three objects was visually more complex than the one with two objects, not surprisingly the baby spent a little longer looking at the picture with three objects.

After the first few trials, however, when the baby had become accustomed to the procedure, a remarkable pattern of behavior began to emerge. The subject spent more time looking at the picture where the number of objects was the same as the number of drumbeats. When two drumbeats were played, the baby looked longer at the picture with two objects. When three drumbeats were played, the child paid more attention to the picture showing three objects.

What was going on? Starkey did not suggest that his subjects had a conscious sense of number. Most likely, what was being observed was a built-in neuronal

response, whereby hearing two drumbeats activates a certain pattern of neuronal activity that makes the brain more receptive to a visual scene showing the same number of objects, two, and likewise in the case of three. But that surely is a forerunner of the abstract sense of number that we develop as we grow older.

So, what is going on here? Many people find math hard, if not impossible to master. In his best-selling book *Innumeracy*[1] (1989), mathematician John Allen Paulos catalogued the many ways that otherwise intelligent and successful people make mistakes with numbers. Yet it seems that we are all born with natural mathematical abilities. Do we somehow lose them as we grow older? Do school math classes somehow manage to drive them out of us? Can we get them back? Even more intriguing, if small babies have inborn mathematical abilities, can other animals also do math?

I started to think about these questions, and others, when I was doing the research for my book *The Math Gene*[2] (2000), and I was amazed at the answers I discovered. Perhaps the most surprising fact is that, far from being an unusual form of thinking that humans have developed and that relatively few can master, mathematics is all around

1. Hill and Wang, 1989.

2. Basic Books, 2000. The book provides answers to the questions: How did the human brain acquire the capacity for doing mathematics? When did it acquire it? and What evolutionary advantage did it confer on our species?

us, sometimes being done by creatures we generally do not credit with much brain power.

In the pages that follow I'll guide you along the same path of discovery that I followed. I guarantee that when you are finished, you will view mathematics in an entirely new light. I'll begin by telling you about some inbuilt mathematical abilities in the animals we are all most familiar with: dogs and cats. Then, after examining impressive capacities for mathematics in various other creatures, we will come back around to people again.

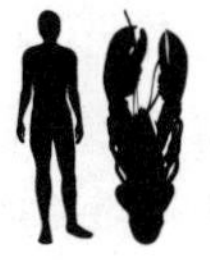

2.

Elvis: The Welsh Corgi Who Knows Calculus

"There's something odd about the way Elvis runs to fetch the ball," Tim Pennings thought to himself one day in 2001. As he did several times a week, Pennings, of Holland, Michigan, had brought his Welsh corgi dog Elvis down to the shore of Lake Michigan to play fetch.

Sometimes Tim would throw the ball along the beach and watch as his dog streaked along the sand to retrieve it. Other times, he hurled the ball out into the water, and that was when he noticed Elvis's curious behavior. If Tim threw the ball straight out into the water, Elvis rushed into the lake and swam directly for it. But if he threw the ball into the water in a diagonal fashion, slanted toward the beach, then instead of simply heading off straight toward the ball, Elvis ran along the water's edge for a while before diving in.

Thousands of dog owners must have seen exactly the same behavior and thought nothing of it. But Pennings is an associate professor of mathematics at Hope College in Michigan, and Elvis's behavior reminded him of a calculus problem he often gave his students to solve. Not just that, but as far as Tim could tell, Elvis was getting the right answer, which was more than he could say for many of his students. "Can my Welsh corgi do calculus?" he wondered.

Tim knew the answer had to be no, but after throwing a diagonal ball into the water a few times and watching the path Elvis took to reach it, he was sure something very interesting was going on. What Elvis seemed to be doing was choosing a path that got him to the ball in the shortest time. But the only way Pennings knew to figure out that path was to use calculus.

Chasing a ball thrown along the sand or else straight out into the lake, the quickest way to reach it is by a straight line directly to the ball. But with a ball thrown out into the lake diagonally, it's much more complicated. Because a dog can run much faster on land than it can swim in water, it is quicker first to run along the water's edge some distance and then dive in and swim the remainder. One way to do this would be to run along the water's edge until directly

opposite the ball, by now bouncing on the surface of the water, and then to make a sharp, right-angle turn into the lake and swim out toward it. But a much quicker way is to run along the beach *partway* toward the point opposite the bobbing ball, and then to swim diagonally from that point in a straight line to the ball. The question is, in order to get to the ball quickest, exactly how far along the beach should Elvis run before jumping into the water?

This is a classic problem that math professors regularly give to their students. Figure 2.1 shows how a college math student is supposed to solve it. The solution requires calculus, a deep mathematical technique discovered by the mathematicians Isaac Newton (1642-1727) and Gottfried Leibniz (1646-1716) in the seventeenth century.

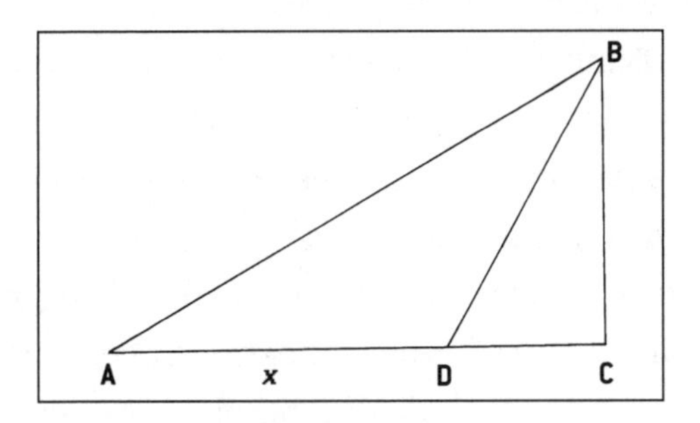

Figure 2.1. The ball-chasing problem. The dog starts at A and the ball is at B. The dog is supposed to run along the beach from A to a point D, and then dive into the water and head directly for B. The problem is to determine the length of the line AD for which the total time taken to reach B is least. To find the answer, you need to know the lengths AC and CB, the speed of the dog running on the sand, and the speed of the dog swimming.

Determined to understand just what Elvis was doing, Pennings set out to collect some data. On his next visit to the beach, he brought the ball, a long tape measure, a stopwatch, and his swimsuit. Time after time—for a total of 35 repetitions in all—Tim hurled the ball, started the stopwatch, raced along the beach after his dog, dropped a marker at the point where Elvis dived into the water, noted the time it had taken Elvis to reach that point, and then followed him in, splaying the tape measure out behind him. Although Tim was left behind running on the beach, he is a strong swimmer, and was generally able to catch up with Elvis before the dog reached the ball, and was able to note the time it took Elvis to swim to the ball. Tim then swam back to the beach, noted the point where he hit land, and then used the tape measure to determine the actual lengths AD and AC. On average, Tim threw the ball 20 meters (65 feet) along the beach and 10 meters (33 feet) into the water. The entire process lasted three hours. Tim stopped when he was exhausted.

Once he got home with his measurements, some easy calculations allowed him to determine all the data required to solve the calculus problem. As he had suspected, Tim discovered that, on average, Elvis took to the water at exactly the right spot given by calculus. The conclusion was inescapable: *In his own way,* Elvis was able to solve a college-level calculus problem.

Tim wrote up his findings and published them in the May 2002 issue of *The College Mathematics Journal,*

published by the Mathematical Association of America. The journal editor published it as the lead article, under the title "Do Dogs Know Calculus?," and put a photograph of Elvis on the front cover—most likely the first time a dog has graced the cover of a mathematical journal.

So how was Elvis doing this? Here is how Pennings explained his findings:

> . . . although he made good choices, Elvis does not know calculus. In fact, he has trouble differentiating even simple polynomials. More seriously, although he does not do the calculations, Elvis's behavior is an example of the uncanny way in which nature (or Nature) often finds optimal solutions. . . . (It could be a consequence of natural selection, which gives a slight but consequential advantage to those animals that exhibit better judgement.)

In other words, Pennings says, the mathematics behind Elvis's remarkable behavior has been calculated by nature. By the process of evolution by natural selection, dogs have developed the ability to do by instinct—perhaps enhanced by experience—exactly what is required to reach the ball in the shortest possible time. In that sense, Elvis is able to solve that one particular calculus problem.

In fact, the dog's mathematical repertoire is more extensive than that. If Pennings had looked more closely

at the way Elvis ran to catch a ball thrown along the beach, he would have noticed another puzzling behavior. Undoubtedly Elvis would not run in a straight line–the way that would get him to the ball soonest–but would follow an arc. Unbeknownst to Pennings, in January of the same year, the *New York Times* reported the results of some other canine research, this time a study of the path a dog would follow when chasing after a Frisbee to catch it.[3] According to the *Times* report, the dog will run in an arc that finishes at the spot where the Frisbee will be when it falls sufficiently close to ground to be caught. Why does the dog do this? Why not run in a straight line, thereby having a greater chance of reaching the Frisbee before it hits the ground?

The question is made the more intriguing by studies of videotapes of baseball fielders running to catch a ball. They too run not in a straight line but in an arc. What is going on here?

The first thing to note is that the calculation required to simultaneously predict where a flying object will land and to compute the direction in which to run to meet it at the right moment is far more complicated than the ball-in-the-water problem Elvis solved. It requires that the ballplayer take account of the speeds of both objects, himself and the projectile. Astronauts face a similar problem when docking a spacecraft with the fast-moving

3. "Fly Ball or Frisbee, Fielder and Dog Do the Same Physics," by Yudhijit Bhattacharjee, January 7, 2003.

international space station. They solve it by using computers to perform advanced mathematical calculations.

Dogs and ballplayers, on the other hand, appear to have unconsciously hit upon a different approach, one that replaces the difficult computation required to simultaneously predict the landing point and determine the path to run, with another computation that, while also difficult, nature has already solved by means of vision. As a result of evolution, dogs and humans can move so as to keep a moving object fixed in their visual field.

In 1995, scientists at Arizona State University suggested that the reason ballplayers follow an arc when running to catch a ball is that they run in a way that makes it look (to the runner) as though the ball is moving in a straight line. In the dog-Frisbee research reported in the *New York Times*, the same scientists attached a small camera and transmitter to the head of a dog in an attempt to capture what the animal was looking at as it raced to catch a flying Frisbee. Like the ballplayer fielding the flying ball, the dog too followed a path that kept the ball's trajectory *looking like* a straight line.

This brilliant strategy provides a dramatic example of how evolution by natural selection can lead to the optimal solution to a problem. In this case, nature's solution was not to equip the animal with a completely new mental algorithm for simultaneously computing a projectile's trajectory and the path to be followed to meet it at the right moment. Instead, nature took advantage of the complicated

mechanisms already in existence to coordinate the visual system and the body-motor system. In order to carry out this strategy the runner must follow an arc rather than a more direct straight line. The advantage is being able to piggyback on some extremely powerful mathematics that nature has embedded in the way the animal's and the ballplayer's visual systems operate. (See chapter 8 for more on the innate mathematics of vision.)

What about cats?

If dogs turn out to be secret mathematicians, what about cats? Do they too exhibit any remarkable calculating abilities? The reluctance of cats to play ball retrieval games, let alone their total refusal to jump into water, means it is not possible to repeat Pennings's experiment with feline pets. Prompted by Pennings's discovery, however, I carried out a search of the literature, and did find one tantalizing lead, in the form of a book called *Calculus for Cats.*[4] The authors are Dr. Jim Loats, a professor of mathematics at Metropolitan State College in Denver, and Kenn Amdahl, a professional writer. Could it be that, as many cat owners maintain, these inscrutable furry creatures have a secret life that most humans are unaware of? Here is how Loats and Amdahl begin their book:

4. Published in 2001 by Clearwater Publishing Company, Inc.

> Approximately four thousand years ago, aliens invaded the earth and began implementing a diabolical plan to enslave humanity, to force us to build their homes, provide them the most expensive and exotic foods, tend their every whim and most trivial wish, regardless of the inconvenience, while they lounged in splendor and did nothing.
>
> These aliens have come to be known as "cats."
>
> The conquest proved simple. Although the creatures lacked opposable thumbs, were inferior in size and had limited capabilities for speech, they had one overwhelmingly superior ability.
>
> They understood calculus.
>
> And humans did not.

Intrigued by the title, and captivated by the introductory passage, I spent some time studying *Calculus for Cats* in order to inform my readers about the mathematical skills of the domestic cat. Sadly, I have to report that Loats and Amdahl's intended readers are almost certainly not cats, but humans. The authors, I decided with great reluctance, had simply hit upon the idea of writing a book ostensibly for cats as a ruse to try to persuade otherwise reluctant university students to master calculus. My suspicions in this direction were initially aroused by the similarity between the book's introductory passage and Douglas Adams's *Hitchhiker's Guide to the Galaxy,* the classic radio series and subsequent book (published 1979)

in which it transpires that the earth and human life upon it was constructed by *mice* as a secret, diabolical plot to serve their own ends. Another major clue to Loats and Amdahl's ulterior intention was the final sentence of the introduction. It reads: "But before you decide that calculus is beyond you, consider this: if cats can learn it, so can you."

Nice try, guys. You have written the world's most entertaining calculus book (perhaps not a particularly remarkable achievement, given the competition). But you wrote it for humans, not for cats. In fact, I do not find a shred of evidence that cats have any mathematical abilities, apart for some innate capacities analogous to Elvis's ball-retrieval skills.

One of the most impressive of those innate skills possessed by cats is their amazing navigational ability. Every few months, a local newspaper somewhere in North America or Europe tells a tale of a domestic cat whose owners move to a new home hundreds or thousands of miles away, whereupon at the first opportunity the home-sick creature sets out and days or weeks later turns up on the doorstep of its old home. Assuming such stories are true, how do they do it? The most plausible possible explanation is that they navigate by the sun, or the stars, or the earth's magnetic field. But as anyone knows who has gone backpacking in the wilderness, it takes some basic trigonometry skills to navigate by using those clues.

Somewhat harder to believe are the stories about cats

that are somehow overlooked and left behind when their owners move house, but which subsequently manage to find their way to their new home. I find these stories implausible because there seems to be no way they could possibly know where their owners have gone. Their sense of smell might enable them to track their owners if the move were made on foot, but when a family piles into a U-Haul truck and trundles along hundreds of miles of interstate highway, it's hard to see how there could be a scent that could be followed–not even if the cat could somehow trot along the freeway pavement with its nose close to the ground.

One particularly dramatic natural-mathematical feat that cats do perform, however, is when they accidentally fall from a wall or tree. On almost every occasion, they manage to orient themselves in the fall so that they land upright on their four paws. Slow-motion film reveals that they do this by rapidly manipulating their body geometry so that gravity–the only substantial force acting upon them–brings them into the upright position. The nearest humans come to performing this remarkable computational feat is when flight control engineers bring under control a satellite that has started to topple or spin about itself while in orbit. This involves some extremely sophisticated mathematics, including the solution (by computer) of a system of partial differential equations involving a dozen variables–a calculation beyond the capacity of most mathematics undergraduates.

Like Tim Pennings's Welsh corgi Elvis, cats too, it seems, have some natural-born mathematical abilities. Moreover, these do not seem to be abilities that our beloved pets pick up from living with us.

In the remainder of this book, we will meet a lot of other creatures (and plants) that evolution has equipped with the ability to perform one or two crucial mathematical tasks. They are nature's mathematicians, and they are all around us, each one performing every day the one particular math problem that ensures its survival. The main lesson we will learn from these examples is that nature (in the form of evolution by natural selection) is arguably the best mathematician of all. Before we go any further, however, let's make sure you understand what professional mathematicians mean by "mathematics."

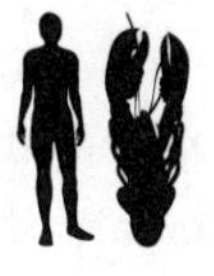

3.

What Is Mathematics?

If you are like most people, it will be obvious to you what it means to do math. Although you might be hard pressed to give a precise definition of mathematics, you will have a general sense of what the subject involves: numbers, arithmetic, algebra, solving equations, geometry, problems about trains leaving stations, proving theorems, etc. You will have no difficulty in saying whether you are any good at it (the answer is generally "no" or perhaps "not very") or whether you like it (again the "no"s are in the majority, although there are more people who answer "yes" than is generally supposed).

But this popular view of mathematics is extremely impoverished, and not particularly representative of the subject as a whole. In particular, although many of the examples I shall describe in this book involve computation with numbers, you will be misled if you automatically think of mathematics as just about, or even mainly about, numbers. Numbers are just one part of one particular kind of mathematics, and as a matter of fact arithmetic computation is not what most mathematicians spend the bulk of their time doing. Nor is the natural mathematics carried out by nonhumans restricted to numbers and arithmetic. Math is about patterns. And patterns are what life is all about.

Numbers arose when our ancestors first recognized that collections of, say, three oxen, three spears, and three women have something in common: threeness. The pattern here is one of numerosity–the size of a collection. Numbers themselves are the objects invented to describe those patterns: the number 1 describes the pattern of oneness, 2 describes twoness, and so on.

Once you have numbers, you can see patterns between those numbers, for example $2 + 3 = 5$, and in this way arithmetic arises. Patterns of shape, important in determining who owns what piece of land or in constructing buildings, give rise to *geometry*, a word that derives from the Greek phrase for "earth measurement." When you combine patterns of shape and patterns of number you get *trigonometry*.

In the 1600s, Isaac Newton in England and Gottfried Leibniz in Germany independently invented *calculus*, the study of the patterns of continuous motion and change. Before calculus, mathematics had been largely restricted to static patterns: counting, measuring, and describing shape. With the introduction of techniques to handle motion and change, mathematicians were able to study the motion of the planets and of falling bodies on earth, the workings of machinery, the flow of liquids, the expansion of gases, physical forces such as magnetism and electricity, flight, the growth of plants and animals, the spread of epidemics, the fluctuation of profits, and so on.

At about the same time that Newton and Leibniz were inventing calculus, the French mathematicians Pierre de Fermat (1601-1665) and Blaise Pascal (1623-1662) exchanged a series of letters in which they developed the beginnings of the branch of mathematics known as *probability theory,* which studies patterns that arise when you repeat a chance event many times, such as tossing a coin or rolling dice. (Their work was motivated entirely by the desires of their wealthy patrons to improve their performance at the gaming tables of Europe.)

Today's computer technology arose from the study of the patterns of logical thought, the branch of mathematics known as *formal logic.*

One distinction that will be important for understanding this book is the difference between mathematics in the conceptual sense and the notations we use to write

it. These days, most mathematics books are filled with symbols. But mathematical notation no more *is* mathematics than musical notation *is* music. A page of sheet music *represents* a piece of music, but the music itself is what you hear when the notes on the page are sung or performed on a musical instrument. The same is true for mathematics; the symbols on a page are a *representation* of the mathematics. When read by someone trained in mathematics, the symbols on the printed page come alive–the mathematics lives and breathes in the mind of the reader.

Without its many symbols, large parts of mathematics simply would not exist. The very *recognition* of abstract concepts and the development of an appropriate language to describe them are really two sides of the same coin. For instance, the use of the numeral "7" to denote the number seven requires that sevenness be recognized as an entity. Having the symbols makes it possible to think about and work with the concept.

This linguistic or conceptual aspect of mathematics is often overlooked, especially in our modern culture, with its emphasis on the procedural, computational aspects of mathematics. Indeed, one often hears the complaint that mathematics would be much easier and more appealing if it weren't for all that abstract notation, which is rather like saying that Shakespeare would be much easier to understand if it were written in simpler language.

When you get beyond the symbols, mathematics, the

science of patterns, boils down to a way of looking at the world—both the physical, biological, and sociological world we inhabit, and the inner world of our minds and thoughts. So far, mathematics' greatest success has undoubtedly been in the physical domain. The Italian astronomer Galileo Galilei (1564-1642) said (with a little paraphrasing), "The great book of nature can be read only by those who know the language in which it was written. And this language is mathematics." In today's age, dominated by information, communications, and computation, there is scarcely any aspect of our lives that is not affected by mathematics; for abstract patterns are the very essence of thought, of communication, of computation, of society, and of life itself.

So can animals really do this stuff?

While my remarks above describe mathematics *as a conscious human endeavor,* in what sense can animals be said to "do math"? Certainly, the paper-and-pencil activity we carry out when we solve a problem is one way of doing mathematics. Generally speaking, it is our way of doing it. But is it the only way?

We would all agree, I think, that when we use a calculator or a computer to solve a math problem, we are still doing mathematics. In many instances we would even be prepared to acknowledge that the calculator or

the computer does the math. What then if some non-human creature solves a similar problem? Is there any justification for denying that a ball-chasing Welsh corgi too is doing mathematics?

You might argue that even the smartest corgi is not consciously aware of doing any calculations. But then, neither is your hand calculator or your computer. You might then counter, "Yes, but the calculator or computer was designed by human engineers to do mathematics." To which I would retort, "But dogs were designed by nature to do (that particular) mathematics."[5]

The mistake in conceiving of mathematics as a purely human endeavor is that we focus almost exclusively on the conscious performance of computational processes—numerical, algebraic, geometric, etc.—often carried out with the aid of a pencil and paper, or, these days, with a calculator or computer. Those kinds of computational activity are certainly part of mathematics, but if you start from the fact that mathematics is about recognizing and manipulating patterns, then viewing the paper-and-pencil calculations we humans do as being all there is to mathematics is like saying that flying is about having wings and

5. I should note that in this book I shall make frequent use of phrases such as "nature designed this" or "nature is efficient," and the like. This should in no way be taken to imply any *intent* on the part of nature, nor indeed that "nature" has any kind of identity. I make no assumptions about the natural world other than (1) stuff happens, and (2) natural selection occupies the driving seat of evolutionary change.

flapping them up and down. Flying is about leaving the ground and moving through the air for extended periods of time. Feathered wings that flap and metal wings engineered by Boeing are just two particular ways of performing that activity.

Once you view mathematics as the science of patterns, and think of doing math as reasoning about patterns, you will find it far less surprising to discover that many living creatures do some mathematics. I'll even give you examples of how plants do math. If you are willing to acknowledge that computers can do math, then there is no justification for denying the same capacities in animals and plants that quite plainly solve problems that we humans solve only by mathematics. After all, on the scale of consciousness, computers lie at the very bottom, well below plants and animals.

This is precisely my starting point. Once you get away from the pencil-and-paper view of mathematics we all get from our school days, and you think about the more fundamental activity that those academic methods provide *just one way of doing,* you find that math is all around us. If you want to find the world's greatest mathematician, you don't need to travel to Harvard or Stanford or Princeton. Just walk into the garden, or take a hike through the forest, or visit the ocean. For nature turns out to be the greatest mathematician of all. Through evolution, nature has endowed many of the animals and plants around us with built-in mathematical abilities that are truly remarkable.

Of course, you have to be a bit careful here. What we need here is not a precise scientific definition but a rule of thumb guided by common sense: In this book, when I speak of mathematics I mean any activity which, when carried out by a human being, would be regarded as doing (or involving) mathematics. For example, Elvis's behavior when retrieving a ball thrown into the lake qualifies as mathematics because the only way we humans have to perform a similar feat is to use math.

If you don't like this definition of mathematics, and cannot bring yourself to call what Elvis does mathematics, then imagine the word "natural" or "nature's" placed in front of "mathematics," and read the remainder of this book as about "natural mathematics" and "nature's mathematicians." Whatever your preferred language, however, if you are interested in the natural world, then I think you will be fascinated (and perhaps surprised) by the first part of this book, where we will look at various creatures that routinely perform feats of natural mathematics. But the story does not end there. In fact, that's where the really intriguing questions arise.

First of all, if many of our fellow creatures can do natural math, surely so can we. What natural mathematical abilities do we have? We have already seen that small babies have some arithmetical ability. What other math problems do we solve all the time–smoothly, effortlessly, and unconsciously, without our knowledge–and how do we solve them? Did our ancestors once possess mathematical

abilities that we have lost as our brains acquired the ability to do things a different way?[6]

Second, what is the difference between our natural mathematical abilities and the math skills we are taught in school? If we possess innate mathematical abilities that are every bit as impressive as those of our fellow creatures, why is it so hard to learn math in school? Why can't we simply tap into those innate abilities we are born with? Or can we? Is it possible to improve the way we teach math by taking a good look at the way we and other living creatures do natural math? Are there parts of mathematics that only a select few among us can hope to master? Or is it just a matter of wanting to?

One place to look for natural mathematical ability, in animals and people, is in the methods creatures use to find their way around. Anyone who has tried to navigate using a map and compass knows that you are unlikely to get where you want to go unless you have some mastery of elementary trigonometry.

To meet a real navigational pro, we begin the next stage of our journey by traveling to the desert sands of North Africa, where we will meet a remarkable mathematician known as Ahmed.

6. I do not answer this last question in this book. Nor can I see how this particular question could ever be answered with any reliability, since the time scale of evolutionary change means we are talking about mental capacities that would have been lost over a hundred thousand years ago, and maybe much further back than that. But I will answer all the other questions, and leave my reader to speculate about long-lost skills of our ancestors.

4.

Where Am I and Where Am I Going?

Ahmed, who was the subject of a research paper published in 1981 by scholars R. Wehner and M. V. Srinivasan, lives in the Tunisian desert, on the northern edge of the Sahara. He has had no formal education, and everything he knows he has picked up by experience. Each day, Ahmed leaves his desert home and travels large distances in search of food. In his hunt, he heads first in one direction, then another, then another. He keeps going until he is successful, whereupon he does something very remarkable. Instead of retracing his steps—

which may have been obliterated by the wind blowing across the sands–he faces directly toward his home and sets off in a straight line, not stopping until he gets there, seemingly knowing in advance, to within a few paces, how far he has to go.

Ahmed has been unable to tell researchers how he performs this remarkable feat of navigation, nor how he acquired this ability. But the only known method is to use a technique known as "dead reckoning." Developed by the ancient mariners of long ago, the method was called "deductive reckoning" by British sailors, who abbreviated the name to "ded. reckoning," a term that in due course acquired an incorrect spelling as "dead reckoning." In using this method, the traveler always moves in straight lines, with occasional sharp turns, keeping constant track of the direction in which he is heading, and keeping track too of his speed and the time that has elapsed since the last change of direction, or since setting off. From knowledge of the speed and the time of travel, the traveler can calculate the exact distance covered in any straight segment of the journey. And by knowing the starting point and the exact direction of travel, it is possible to calculate the exact position at the end of each segment.

Dead reckoning requires the accurate use of arithmetic and trigonometry, reliable ways to measure speed, time and direction, and good record-keeping. When seamen navigated by dead reckoning they used charts, tables, various measuring instruments, and a considerable

amount of mathematics. (The main impetus to develop accurate clocks came from the needs of sailors who used dead reckoning to navigate vast tracts of featureless ocean.) Until the arrival of navigation by the Global Positioning System (GPS) in the mid 1970s, sailors and airline pilots used dead reckoning to navigate the globe, and in the 1960s and 1970s, NASA's Apollo astronauts used dead reckoning to find their way to the moon and back.

Yet Ahmed has none of the aids that mariners and lunar astronauts made use of. How does he do it? Clearly, this particular Tunisian is a remarkable individual. Remarkable indeed. For Ahmed measures little more than half a centimeter in length. He is not a person but an ant—a Tunisian desert ant, to be precise. Every day, this tiny creature wanders across the desert sands for a distance of up to fifty meters (165 feet) until it stumbles across the remains of a dead insect, whereupon it bites off a piece and takes it directly back to its nest, a hole no more than one millimeter in diameter. How does he navigate?

Many kinds of ant find their way to their destination by following scents and chemical trails laid down by themselves or by other members of the colony. Not so the Tunisian desert ant. Observations carried out by Wehner and Srinivasan, the researchers mentioned above, leave little room for doubt. The only way Ahmed can perform this daily feat is by using dead reckoning.

Wehner and Srinivasan found that, if they moved one

of these desert ants immediately after it had found its food, it would head off in exactly the direction it *should* have taken to find its nest if it had not been moved, and, moreover, when it had covered the precise distance that should have brought it back home, it would stop and start a bewildered search for its nest. In other words, it knew the precise direction in which it should head in order to return home, and exactly how far in that direction it should travel, even though that straight-line path was nothing like the apparently random zigzag it had followed in its search for food.

A recent study[7] has shown that the desert ant measures distance by counting steps. It "knows" the length of an individual step, so it can calculate the distance traveled in any straight-line direction by multiplying that distance by the total number of steps.

Of course, no one is suggesting that this tiny creature is carrying out multiplications the way a human would, or that it finds its way by going through exactly the same mental processes that, say, Neil Armstrong did on his way to the moon in *Apollo 11.* Like all human navigators, the Apollo astronauts had to go to school to learn how to operate the relevant equipment and how to perform the necessary computations. The Tunisian desert ant simply

7. S. Wohlgemuth, B. Ronacher, and R. Wehner, "Odometry in Desert Ants: Coping with the Third Dimension." *Journal of Experimental Biology,* forthcoming.

does what comes naturally—it follows its instincts, which are the result of hundreds of thousands of years of evolution.

In terms of today's computer technology, evolution has provided Ahmed with a brain that amounts to a highly sophisticated, highly specific computer, honed over many generations to perform precisely the measurements and computations necessary to navigate by dead reckoning. Ahmed no more has to *think* about any of those measurements or computations than we have to think about the measurements and computations required to control our muscles in order to run or jump. In fact, in Ahmed's case, it is not at all clear that he is capable of anything we would normally call conscious mental activity.

But just because something comes naturally or without conscious awareness does not mean it is trivial. After all, almost fifty years of intensive research in computer science and engineering has failed to produce a robot that can walk as well as a toddler can manage a few days after taking its first faltering steps. Instead, what all that research has shown is how complicated are the mathematics and the engineering required to achieve that feat. Few adults ever master that level of consciously performed mathematics—let alone a small child that runs with perfect bodily control for the candy aisle in the supermarket. Rather, the ability to carry out the required computations for walking comes, as it were, hardwired in the human brain.

So too with the Tunisian desert ant. Its tiny brain might have a very limited repertoire. It may well be incapable of learning anything new, or of reflecting consciously on its own existence. But one thing it can do extremely well–indeed far better than the unaided human brain, as far as we know–is carry out the particular mathematical computation we call dead reckoning. That ability does not make the desert ant a "mathematician," of course, but that one computation is enough to ensure the desert ant's survival. And that is precisely how evolution by natural selection works.

Nature performed a similar trick with another creature we don't normally think of as smart: the lobster.

LOBSTER, ANYONE?

I've known people who refuse to eat meat because we kill animals to obtain it, but who nevertheless are happy to eat seafood. High on the list, for some of them, is a delicious Maine lobster. After all, just look at it: can you imagine anything more primitive, anything less likely to have a conscious sense of its own existence? Next time you sit down for a lobster dinner, however, ponder this: You will be tucking in to one of nature's most accomplished navigators. For the fact is, the common lobster has a geographical location system that humans can match only with the latest, most sophisticated version of the GPS, the

hugely expensive navigation system (initiated in 1974, operational in 1994) that depends upon satellites that orbit the earth, the most accurate timekeeping devices ever devised, and masses of computer power and advanced mathematics.

What humans accomplish with mathematics and technology, the lobster achieves by being able to "see" the earth's magnetic field, and not merely in the sense of detecting the magnetic poles–the lobster's system is much more sophisticated than that. The earth's magnetic field varies from one place to another, in direction, angle to the earth, and intensity. The lobster appears to be able to use this variation to determine exactly where it is. This was discovered only a few years ago, by ocean scientist Ken Lohmann of the University of North Carolina and his Ph.D. graduate student Larry Boles.[8]

It took Boles six years of study of the Caribbean spiny lobster in the waters near the Florida Keys before he was convinced that they possess this amazing ability. To demonstrate the fact, he tried all kinds of ruses to confuse them. He removed them from the ocean and put them in a lightproof plastic container, drove them around in circles in his boat, took them ashore and drove them around in the back of his pickup truck, placed them next to powerful magnets to distort the earth's magnetic field, and then

8. Larry C. Boles, Kenneth J. Lohmann, "True Navigation and Magnetic Maps in Spiny Lobsters," *Nature* 421, 2 January 2003, pp. 60–63.

dropped them back in the ocean in a new location. As soon as they were released the lobsters headed off directly toward their home. They did so even when Boles placed rubber caps over their eyes, so they were not navigating by light. But to be doubly sure, Boles put some lobsters into a marine tank in his lab and subjected them to an artificial magnetic field that mimicked that of the earth. The lobsters headed off in exactly the direction they would have had to follow to get home if the field had been the earth's natural one.

The researchers suspect that the lobster's navigational ability may make use of small particles of magnetite, an iron oxide, located in two masses of nerve tissue toward the front of the creature's body.

The Mathematical Secrets of Migration

Moving from the oceans to the skies, birds provide another example of remarkable navigational ability. Every year, millions of birds migrate thousands of miles to and from their winter home. How do they know which direction to fly? There are several possibilities, but most of them seem to require mathematical computations that most humans would find challenging. How do they do it?

To put the question another way, why is it that a pilot of a Boeing 747 needs a small battery of maps, computers,

radar, radio beacons, and navigation signals from GPS satellites, all heavily dependent on masses of sophisticated mathematics, to do what a small bird can do with ease, to fly from point A to point B?

To give you some idea of the distances that can be involved: arctic terns fly an annual round-trip that can be as long as 22,000 miles, from the Arctic to the Antarctic and back. On the trip south, they make a regular stopover on the Bay of Fundy, fly a grueling, three-day nonstop leg across the featureless North Atlantic, and make their way along the entire west coast of Africa. They return by a different route, coming up the east coast of South America and North America. Other seabirds also make amazingly long trips: the long-tailed jaeger flies 5,000 to 9,000 miles in each direction, the sandhill and whooping cranes are both capable of migrating up to 2,500 miles per year, and the barn swallow logs more than 6,000 miles annually.

Some of those globe-hopping birds can also fly high. Bar-headed geese have been spotted flying across the Himalayas at 29,000 feet. Other species seen above 20,000 feet include the whooper swan, the bar-tailed godwit, and the mallard duck. From radar studies, scientists have been able to conclude that, like long-haul airline pilots who adjust their cruising height to avoid a strong headwind or to ride the jet stream, birds change altitudes to find the best wind conditions. To avoid battling against a headwind, most birds stay low, where ridges, trees and

buildings slow the wind. To ride a tailwind, they get up high where the wind is as fast as possible.

So how do they find their way? Scientists still have a long way to go before they understand completely how birds navigate, but the available evidence suggests that they use a combination of different methods.

First, birds may use visual clues. Many animals learn to recognize their surroundings to determine their way. They remember the shape of mountain ridges, coastlines, or other topographic features on their route, where the rivers and streams lie, and any prominent objects that point to their destination. Birds may use this method to locate their nest, but it seems unlikely that it will support flights over long distances. And it clearly cannot be used for navigating over large bodies of water or for flying at night, both of which are done by many species of bird every year.

Other methods depend on determining the direction of the North Pole. Humans do this using a compass or by the position of the sun in the sky. But as any sailor or mountain walker will confirm, knowing where north lies is only part of what it takes to navigate. You also need to know the direction in which to travel relative to north. To do that, humans require, in addition to the compass or the sun, a map together with some arithmetic, geometry, and trigonometry.

So, how do birds do it? Let's start with the problem of orientation: How do the birds know which direction is north? One possibility for setting direction is to use the position of the sun in the sky. Many birds—and other creatures such as

the honeybee–have been shown to use the sun to determine where north is. This is not as simple as it might first appear, since the sun changes its position in the sky throughout the day and the pattern of those daily changes itself varies with the seasons of the year. To use the sun to set the direction to north, you have to know where the sun is located in the sky at each time of the day at the precise time of year the journey takes place. For a human navigator, that task alone requires mastery of trigonometry, in addition to the mathematics required to plot a course from the position of the sun in the sky.

One obvious problem with birds using the sun to navigate is, what do they do at night? And what do they do during the day if it is very cloudy? Since many birds fly at night and in cloudy weather, navigating by the sun is clearly not the only method they use. One possibility for nocturnal travel is to be able to detect the polarization of the light from the moon. Although this light is considerably weaker than sunlight, given the appropriate detection apparatus, this is a possibility. One creature that we know for sure makes use of this navigational aid is the dung beetle. Writing in the journal *Nature* in 2003,[9] a group of researchers from Sweden and South Africa explained how dung beetles use polarized light from the

9. Marie Dacke, Dan-Eric Nilsson, Clarke H. Scholtz, Marcus Byrne and Eric J. Warrant, "Animal Behaviour: Insect Orientation to Polarized Moonlight," *Nature* 424, 3 July 2003, p. 33.

moon to orient themselves when they travel at night. Put a polarizing filter between the moon and a dung beetle, the researchers found, and at once the poor creature becomes hopelessly confused, and will start to wander around in circles, when just a few moments earlier it had been headed determinedly and accurately back toward the pile of dung whose location it knew from an earlier visit. But, of course, navigation by moonlight is only possible on cloudless nights when there is a moon.

Another way to determine orientation, which works at night as well as by day, whether cloudy or not, is to make use of the earth's magnetic field. This, of course, is exactly what we do when we use a magnetic compass. Some birds use a similar method to navigate. For instance, inside the skull of a homing pigeon is a small pod of magnetic particles, which provides the bird with a tiny magnetic compass in its head. By attaching small magnets to the heads of test birds, researchers have shown that homing pigeons navigate by means of the earth's magnetic field. The magnets deflect the earth's magnetic field around the birds, and cause them to fly off course. (Put simply, with the magnet attached to its head, the bird thinks that any direction it is facing is north.)

Star navigation provides another means of navigating at night. This method was used by human sailors in times past. At least one species of bird, the indigo bunting, is known for sure to use the stars to navigate, and it is generally believed that all birds do so. It appears that they

learn to recognize the pattern of stars in the night sky when they are still fledglings in the nest. A few years ago, a study found that nestling indigo buntings in the northern hemisphere watch as the stars in the night sky wheel around Polaris, the north star, which lies due north for those in the Northern Hemisphere. Scientists speculated that being able to identify Polaris in the night sky could help birds identify north. To test this hypothesis, they showed the birds a natural sky pattern inside a planetarium. The birds flew in a direction consistent with being able to detect the motion of the stars. When the experimenters changed the setup so that Betelgeuse was now the star around which the stars rotated, the birds flew in a direction consistent with Betelgeuse being the pole star. They no longer went where they should have relative to Polaris. Thus, they were not using the locations of specific star patterns; they were noticing which star the others rotated around. In other words, it was not the star patterns, but how the stars moved that counted. For the birds, "north" was where there was a star around which all other stars moved.

In fact, birds seem to make use of their ability to read the stars in order to solve a problem with using the earth's magnetic field to navigate: recalibration of their built-in compass. Magnetic north is 1,600 kilometers from the North Pole, and that means that migrants leaving northern Alaska and following magnetic south would be traveling due west! Birds have to constantly recalibrate

their magnetic compasses as they travel over long distances. One way they do this is by comparing their internal compass setting against their star navigation during their rest stops along the migration route. (If they don't have enough time to complete the recalibration at a rest, they get lost.)

An alternative recalibration technique was discovered in 2004 when a team of researchers showed that certain thrushes check their bearing each evening against the direction of a setting sun.[10] To verify this theory, the researchers fastened small radio transmitters to the birds so they could follow them on the ground, and in the evening, as the sun was setting, they subjected the birds to a magnetic field different from the earth's natural magnetic field, and strong enough to override it. Sure enough, the next morning, the birds headed off in the wrong direction–the heading that would have been correct if the imposed field had been in the same direction as the earth's. The next evening, the birds were able to carry out the calibration without interference, and the following morning they resumed their journey on the correct course.

Whatever the method, for long-distance travelers like migrating birds, simply setting the internal compass is sufficiently tricky to challenge a college math graduate.

10. William Cochran, Henrik Mouritsen, and Martin Wiselski, "Migrating Songbirds Recalibrate Their Magnetic Compass Daily from Twilight Clues," *Science*, Vol. 304, 16 April 2004, pp. 405-407.

For that matter, reading the stars also has its complications: The pattern of stars in the sky keeps changing as the birds travel north or south, with new constellations constantly appearing on the horizon. Adjusting for the change in the sky is another thing that humans can do accurately only by using mathematics.

One additional navigational possibility is that birds discern polarization patterns in sunlight. As the sun's rays pass though our atmosphere, tiny molecules of air allow light waves traveling in certain directions to pass through, but they absorb others, causing the light to be polarized. We can see the polarization effect if we look up into the sky at sunset. The polarized light forms an image like a large bow tie located directly overhead, pointing north and south. It seems that some birds can detect the gradation in polarization from the nearly unpolarized light in the direction of the sun to the almost 100 percent polarized light 90 degrees away from the sun, and this provides them with a giant compass in the sky. Honeybees also appear to use the polarized light to find their way on cloudy days, when the sun can't be seen. All they need is a small patch of blue sky to see the sun's rays through, and the polarization effect shows them the way.

Whatever method the birds use to orient themselves, however, orientatation is just part of navigation. For humans, at least, setting the right course from the orientation requires trigonometry. How do the birds do it?

Birds are not the only animals that are able to orient

themselves. Many sea creatures also migrate over large distances. For instance, salmon have a seasonal migration that can take them across thousands of miles of ocean whose surface features look the same mile after mile. Studies have shown that they navigate primarily by the position of the sun during the day and by the stars at night. When the sun is not present and the stars are obscured by clouds, they use the earth's magnetic field. To demonstrate this fact, researchers put salmon into a large tank around which were arranged a number of electromagnets that could be used to change the direction of the magnetic field (overriding the natural magnetic field). When the sun was visible, changing the magnetic field from the prevailing North-South to an artificial East-West did not cause the salmon to change direction; they continued to swim south. But when the sky was cloudy, changing the magnetic field to East-West caused the fish to swim toward the artificial magnetic south. Similar experiments have shown that whales and sea turtles also navigate by a combination of observations of the sky and the earth's magnetic field.

And then there is the amazing North American spectacle provided every year by the monarch butterfly. This bright orange insect is a familiar sight in gardens all across the United States and Canada throughout the summer months. Then, every September, all one hundred million or so of them set out on a two-and-a-half-month trek to their winter homing ground, a single, thirty-acre area of

mountain pines in the Mexican state of Michoacán, west of Mexico City. None of these fall migrants has ever been there before. They are the third- or fourth-generation descendants of the long-dead ancestors who made the long journey north in the spring. And yet the majority of them manage to find their way to the species's traditional winter habitat, as far as two thousand miles away. Only recently have scientists begun to understand how they perform this seemingly miraculous feat.

We know that they navigate primarily by means of the sun. We know too that they are sensitive to ultraviolet light, so they are not dependent on a cloudless sky. (Monarchs flying in full sunshine in an enclosure will stop immediately if a filter is applied that blocks the sun's ultraviolet waves.) But orientation by the sun requires knowing the time of day, so it was long presumed they must make use of some form of inner clock. This was finally demonstrated in the spring of 2003 by a team led by Steven M. Reppert of the University of Massachusetts Medical School.[11] Like most creatures, the monarch butterfly regulates its daily activities by means of what is known as its circadian clock. (The term "circadian" comes from the Latin *circa diem,* meaning "about a day.") This naturally occurring biological clock tends to be fairly accurate over the short term but requires periodic recalibration to take account of the changing length of day and

11. Steven M. Reppert et al., *Science,* Vol. 300, 23 May 2003, pp. 1303-1305.

night during the course of the year. (Disruption of the human circadian clock is what causes jet lag when we fly long distances.)

To test for the role played by the circadian clock in the monarch, Reppert and his team put a group of the insects into a laboratory chamber where, for a whole week, they were subjected to a light pattern typical of early September, twelve hours of daylight from 7:00 A.M. to 7:00 P.M., followed by twelve hours of darkness. When released in the morning, they headed off with the sun over their left "shoulders," the direction they would have to go to get to Mexico at that time of year. But another group of monarchs, whose week-long period of artificial daylight had been on a 1:00 A.M. to 1:00 P.M. cycle, oriented themselves with the morning sun over their right "shoulders," the correct strategy to adopt to head for Mexico if in fact it had been late afternoon. Still a third group of monarchs were subjected to a week of continuous light. This broke their circadian cycle, and as a result, when the butterflies were released they simply flew directly in the direction of the sun.

Of course, for monarchs as for birds, positioning of the sun relative to the time of year and the time of day, even with the seeming accuracy they bring to the task, solves only part of the problem of navigation. For monarchs too there remain the directional calculations that humans solve by using trigonometry. As University of Kansas entomologist Orley "Chip" Taylor (a member of the monarch research team) has observed, "Navigation implies directed

flight toward an unseen goal. A monarch in Georgia will be flying 270 degrees [due west] to get to Mexico, while a monarch at the same latitude in Texas will fly 220 degrees [southwest]. You tell me how that happens."

Birds, salmon, whales, sea turtles, monarch butterflies, lobsters, even dung beetles–nature has equipped these and other migrating creatures with the ability to find their way accurately, often across thousands of miles of the earth's surface. Internal magnets, the ability to "read" the earth's magnetic field, and eyes that can detect polarized or ultraviolet light are only a part of the story. Accurate navigation also requires a brain that can process the position and orientation information, and then combine it with the time of year and an internal daily clock in order to set the direction to be flown, walked, crawled, or swum at each moment of the journey.

The Tunisian desert ant, the lobster, and other creatures all simply follow their instincts, but when we try to understand how they do it, we have to resort to mathematics. The only way we can describe the accomplishment of a migrating bird or fish is to say that it has a brain that has evolved to carry out the trigonometrical calculations necessary to determine north from the position of the sun or to set a course based on a knowledge of where the North Pole lies. Since human brains are not equipped with the same built-in capabilities (or if we are, we are not aware of them), human navigators cannot do it the same way. We have no alternative other than to "do the math" in order to

find our way around—or else, these days, to make use of equipment that has been designed and built to do the calculations for us.

The term "bird brain" may under normal circumstances be an appropriate metaphor to refer to someone with poor intellectual ability. But when it comes to navigation, birds undoubtedly leave us in their wake. They and the other migrating creatures join the Tunisian desert ant as accomplished members of nature's club of natural mathematicians. Among the other members are two creatures that also use mathematics to determine which way to go: bats and owls. But in this case they use the math for a quite different purpose: to kill—a function they perform with as much efficiency as a guided missile.

BAT MOBILE

Figure 4.1. The familiar brown bat, found all over North America.

How much do you know about bats? Say whether each of the following statements is true or false.

1. Bats are not birds; they are nature's only flying mammal.
2. Bats are found all over the world, except in the extreme desert and polar regions.
3. Over one thousand species of bat are known.
4. Bat wings are really highly maneuverable hands, with long fingers that are connected by a thin membrane of skin.
5. The bat can manipulate its wings to trap an insect.
6. The bat can hover motionless in the air like a hummingbird.
7. Bats are one of nature's most efficient insect control devices. The tiny brown bat found all over North America can eat up to seven thousand mosquitoes a night.
8. Contrary to the popular saying "as blind as a bat," bats have excellent eyesight.
9. The oft-repeated claim that bats will mistake a woman's hair for prey and become entangled in it is a misconception.
10. The belief that the so-called vampire bat will suck human blood is a misconception.
11. Bats use a sonar system to navigate and catch prey at night that is far more accurate than

anything human engineers have produced. The U.S. Navy has tried to imitate the system to develop better minesweeping technology.
12. Using its sonar system, on a pitch-black night a bat can swoop down and catch a beetle in midair as it rises from the grass.
13. Bats seek their prey both in open terrain and in the trees and the undergrowth.
14. Some robotics engineers have been so impressed with bat sonar that they have used sonar detection instead of cameras to guide robotic devices, basing their design on studies of bats.
15. Bats are among one of nature's most impressive mathematicians.

All fifteen statements are correct. I will briefly comment on a few of them.

First, bats are indeed mammals: they have teeth and a body covered in fur; they give birth to live young, which they nurse with milk.

"Blind as a bat"? Hardly. Statement 8 is correct: Bats have excellent eyesight. They use it for long-distance navigation during daylight hours. But they are essentially nocturnal creatures—that's when they do their hunting for food—and for night flying they rely on their sonar system, a system that is so miraculous, it is hardly surprising that the false belief has developed that bats are blind.

The misconception in statement 9 probably arises from bats' swooping down to catch insects that hover around a woman's head, attracted by her perfume or the scent of her hairspray. A bat's sonar-based navigation system is so accurate, however, that it can easily tell that a human being is too big to constitute a potential meal. Although they might come in quite close, it is the insects they are after, and with their pinpoint sonar they are unlikely to come into contact with a woman's hair, let alone become entangled in it.

The belief in statement 10 is the stuff of Hollywood movies, but what is the reality? It is true that most bats are carnivorous, but the largest prey any species will attack are frogs, lizards, birds, small mammals, and fish. There are also some vegetarian species of bat that live off fruits, nectar, and pollen. While most bats are fairly small, the largest of all, the so-called flying fox, weighs up to a kilogram and can have a wingspan of two meters. But far from being a Hollywood monster, it eats only fruit, not blood. As for that infamous vampire bat, which lives in Central and South America, yes, it really does suck blood, but it is fairly small and only attacks birds and small mammals, not people.

Finally, regarding statement 13, the misconception that bats seek their prey only in open terrain, probably arose because, without modern night vision equipment, it is virtually impossible to observe bats in densely vegetated terrain. Thus, all early studies were carried out in the open, where the bats could be seen silhouetted against the moonlit sky. In fact, a bat can fly into a bush at night to

capture its prey, its sonar system being able to pinpoint the echo from the prey among the echoes from all the branches and leaves of the bush.

Bat sonar is truly one of nature's marvels. As long ago as the eighteenth century, people speculated that bats "see" with their ears, which are indeed pretty large. But it was much more recently that the precise mechanism was determined. Bats emit high-frequency sounds, high-pitched beeps or clicks, beyond the range the human ear can detect, and they listen for the echoes as they bounce off objects. It is a highly efficient mechanism that enables them to navigate safely at night, often with high speed, avoiding obstacles (including other speeding bats) and catching insects in flight.

Although we cannot know what it feels like to be a bat, possibly the best way to understand this system of *echolocation* is that the echoes create a sonar image of the environment, analogous to the visual image we obtain from the light entering our eyes.

Echolocation is not entirely the same as vision, however. There are some important differences. As we shall see in chapter 8, when we see, light from an object enters our eyes and creates a two-dimensional image on the retina, which our brain then interprets as a three-dimensional visual image. The light from the different parts of the visual field all enters the eyes at the same time, of course, but different light rays have varying wavelengths and intensities, which the eyes can perceive in order for the

brain to construct its internal visual image. When a bat emits an ultrasonic wave, the wave bounces back from any object it meets. The more distant the object, the longer it takes the echo to return. Thus, unlike light, the incoming sonar waves come in spread over time. It is in large part the arrival-time differences between the echoes that the bat's brain uses to create an internal "sonar image" of the world in front of it.

One difference between light vision and sonar perception is that the eye has a lens that focuses the incoming light rays. Our visual image is primarily in the left-to-right and up-and-down direction, with depth being created in the brain from a variety of clues in the incoming light signal. Sonar vision, in contrast, does not have the equivalent of a lens, and the sonar image is primarily front-to-back, built up largely from time differences between the returning sound waves.

Marine scientists use echolocation to map the surface of the ocean floor, using a loudspeaker mounted on the underside of a ship's hull to send electronically generated pings down to the sea bottom, and timing the delay to the echo at each point as the ship moves slowly along the surface. Sophisticated computer equipment converts the echo timings into a contour map of the seabed. The software that converts the echo delays to a contour map uses mathematics. In its own terms, the bat's brain must perform the same calculations to produce the sonar image that it uses to navigate and catch prey.

How good can echolocation be? Are those claims in statements 11, 12, and 13 really true? Admittedly, sonar technology for ocean floor mapping is not particularly accurate. For one thing, the ship is constantly bobbing up and down on the surface. But it is adequate for drawing contour maps of the seabed. Sonar developed for military use and for other scientific purposes, however, can be far more accurate. But not as accurate as the bat. Scientists have studied bats in the laboratory, and found that they can perceive and distinguish overlapping echoes separated by just two millionths of a second. This enables them to discriminate between objects just three-tenths of a millimeter apart—about the width of a pen line on a sheet of paper. That's a lot better than we can do with vision, and between two and three times more accurate than the best human-made sonar equipment.

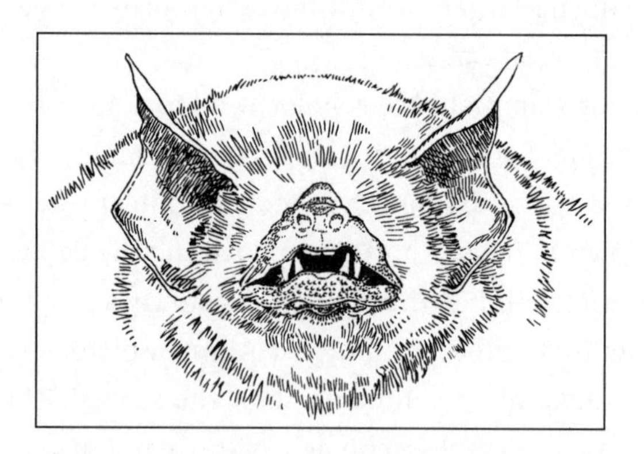

Figure 4.2. The mustached bat. Its sonar-based detection system allows it to swoop on its target with more accuracy than the most advanced jet fighter aircraft.

Mustached bats, named for their facial appearance (see figure 4.2), are an interesting and frequently studied species. They catch flying insects and other moving prey using a particularly sophisticated form of echolocation in which the sound they emit has two phases.

The first phase is a sound of constant frequency, which not only enables them to form an accurate contour map of the terrain, but also allows them to determine the motion and speed of insects or other prey, using the familiar Doppler shift effect. This is where the returning sound wave from an approaching object is compressed, causing its frequency to rise, and the returning wave from a receding object is elongated, causing its frequency to fall. We notice this when a police car passes us in the street with its siren blowing: The siren rises in pitch as the car approaches, and then falls after the car passes by and starts to recede. Astronomers use the Doppler shift in light waves to determine the rate at which distant stars and galaxies are moving away from us.

The second phase of the pulse has variable frequency, and enables the mustached bat to make a highly accurate determination of the distance to an object as well as some of its finer details.

This highly sophisticated echolocation system—sometimes called CF-FM, for "constant frequency-frequency modulated" echolocation—enables the bat to thrive in densely vegetated regions, where it is able to pick out insects moving among all the vegetation.

When the mustached bat produces its CF-FM pulse, it also produces harmonics. Since evolution generally follows some purpose, it is reasonable to assume that the bat does this in order to extract additional information from those harmonics. Laboratory experiments have shown that a significant portion of the mustached bat's brain is devoted to processing the frequency range of the second harmonic. Since this frequency provides a particularly accurate Doppler shift, some scientists have speculated that this is precisely why the harmonics are generated.

Besides the impressive natural audio sensing and audio signal processing technology these bats are using, they are also into some heavy-duty mathematics. As anyone knows who has looked at how astronomers go from a measurement of the Doppler shift to a determination of the speed, it requires some fairly advanced mathematics. And the Doppler shift calculation is just one part of the picture. Flying at high speed at night, the mustached bat can do everything the best trained "Top Gun" fighter pilot can do, and more. It can determine the terrain, pick out moving prey, calculate the direction and speed in which the prey is moving, adjust its own speed and flight path, predict where the prey will be at contact time, and secure a capture. In aeronautical terms, the bat can discern terrain contour and distance, relative velocity of a target, flutter of target, range, size, fine characteristics, azimuth, and elevation. That makes it far more efficient than any multimillion-dollar guided missile that the

U.S. military has been able to put into the field, and more than a match for any highly trained combat pilot in a billion-dollar jet fighter.

EYES TO KILL

Figure 4.3. The owl. Wise sage or lethal killing machine?

Turning now to owls, while bats have a frightening public image, owls are seen as peaceful, wise sages. It's all an illusion. The owl's large, staring eyes, topped by that prominent brow, might give us the impression of thoughtfulness,

but the owl is a highly engineered killing machine with a precision guidance system that, like the bat's, depends on a sophisticated built-in calculating engine.

The first thing you notice when you see an owl are its eyes. They're huge. Take the great horned owl, for example. (This is one of some 140 different species, and is found all over North America.) If this creature were the same size as a human, its eyes would be the size of oranges. Those eyes can take in so much light that the owl can see in extremely dark conditions. And then there's that stare. The owl's eyes are fixed in their sockets–the owl can move them neither from side to side nor up and down. Instead, the creature moves its entire head in order to see; its neck is capable of turning the head through 270 degrees–three-quarters of a complete circle.

The owl uses its eyes and its ears to detect danger and to spot prey, which may include rabbits, mice, squirrels, shrews, weasels, gophers, frogs, snakes, bats, beetles, grasshoppers, scorpions, birds, ducks, grouse, pheasants, other owls, and occasionally domestic cats. Once the prey is detected, however, the eyes play little role in what happens next. What makes the owl such an efficient hunter is its incredible hearing. Unlike the bat, which sends out sounds and listens for the echoes, the owl swoops in on its prey guided by the tiny rustles and other barely perceptible sounds coming from the luckless victim.

The owl's hearing is extremely acute. Those heavy eyebrows that help to give the owl that thoughtful appearance

are actually part of a face that is contoured to direct sound into the ears, much like the parabolic dish of an astronomical radio telescope. Take a closer look at the ears themselves, and you'll see that they are not placed symmetrically on the head, as human ears are, but are offset, with the right ear generally larger (often 50 percent larger than the left) and set higher up the skull. This arrangement makes the owl's hearing an awesomely accurate nighttime guidance system.

The owl pinpoints the position of its prey by the mathematical technique known as triangulation. Given the angle between the two ears when both are focused on the same sound source, a trigonometrical calculation will determine how far away the source is. As we shall see in chapter 8, our brains perform the same calculation to determine distance from the angle between our eyes. (In both cases, this computation is a natural one, of course, built into the creature by evolution.) The accuracy the owl achieves by using its ears in this way is every bit as good as the best-trained tennis player can achieve with vision.

The asymmetry of the ears enables the owl to determine the motion of its prey. The sound signal it picks up in one ear is the mirror image of that coming into the other, except that the right one is stronger (the ear is bigger) and displaced 10 to 15 degrees higher. The owl uses this difference in the two perceived signals to determine the prey's motion. The math required to perform this computation would challenge most high school math whizzes.

The way nature has equipped the owl to do this math is that it turns its head to equalize the sounds coming into both ears, and the amount of turn is exactly what is required to keep the moving prey dead ahead.

The poor prey is unlikely to know what hit it. The owl is a marvel of aerodynamic engineering, its feathers being extremely soft to enable it to fly in almost total silence. The feathers at the front edges of the owl's wing are saw-toothed, which makes the wind pass over the wings noiselessly, so there is not even the sound of rushing air to warn the prey of the impending attack.

One of the owl's four toes is reversible, allowing each foot to form a two-two pincer grip to catch its prey as it swoops down on it. Once the capture has been made, the owl swallows small animals whole, and tears larger ones to pieces.

Like the "blind" bat, the "wise" owl is a precision-built hunter/killer that uses a sophisticated natural guidance system, depending on advanced natural mathematics, to go about its business with astonishing accuracy. Yet again we see that nature has equipped some of our fellow creatures with sophisticated capacities that enable them to perform acts that we can do only after spending many years learning mathematics, or else by relying on technology that in turn depends on the prior mathematical efforts of others.

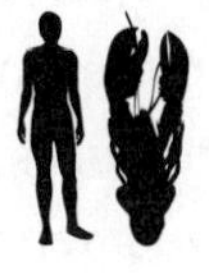

5.

Nature's Architects: The Creatures That Can Do the Math of Construction

Let's leave navigation for a while, and turn our attention to construction. Anyone who has built a house or had an extension added to their home knows that the first step is to draw up plans—precisely drawn scaled drawings that tell the builder what goes where. Drawing accurate blueprints for house building requires basic trigonometry—the measurement of lengths and angles—and so too does the subsequent construction work. Without the precision afforded by trigonometry, the result of a house construction project could easily turn out to be a disaster.

But humans are not the only creatures that erect buildings. Among the structures created by various creatures, when it comes to geometric elegance, surely none surpasses the beautiful, repeating, six-sided shape of the honeycomb. (See figure 5.1.) Honeybees create these great works of architectural art to store the honey they manufacture.

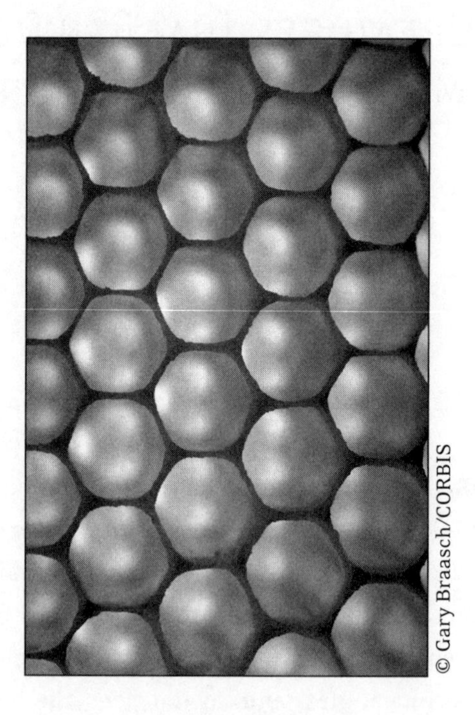

Figure 5.1. The honeycomb. The precision with which bees construct this optimally efficient storage device would satisfy any civil engineer.

Starting with the fourth-century Greek geometer Pappus, many people suspected that the elegant, hexagonal (six-sided) shape of the honeycomb was a result not so much of bees' innate sense of geometric beauty but rather was yet another manifestation of nature's efficiency at work. The common belief was that the repeating hexagonal pattern you see in a cross-section of a honeycomb was the architecture that used the least amount of wax to build the walls.

Pappus himself suggested this hypothesis, which became known as the Honeycomb Conjecture, in an essay on what he called "the sagacity of bees." The hypothesis resisted all attempts to prove it until 1999, when mathematician Thomas Hales of the University of Michigan announced that he had finally managed to crack the puzzle.

It was not until the advent of close-up film techniques that scientists knew for certain how bees build their honey stores. It is an impressive feat of high-precision engineering. Young worker bees excrete slivers of warm wax, each about the size of a pinhead. Other workers take the freshly produced slivers and carefully position them to form vertical, six-sided, cylindrical chambers (or cells). Each wax partition is less than one-tenth of a millimeter thick, accurate to a tolerance of two-thousandths of a millimeter. Each of the six walls is exactly the same width, and the walls meet at an angle of precisely 120 degrees, producing a cross section that is what mathematicians

call a regular hexagon, one of the "perfect figures" of geometry.

Why do bees chose the hexagonal cross section? Why don't they make each cell triangular, or square, or some other shape? Why have straight sides in the first place? After all, the warm wax could just as well be formed into curved walls. Although a honeycomb is a three-dimensional object, because the individual cells are all cylindrical, the total area of the wax walls depends solely on the shape and size of the cross section of the cells. Thus, the mathematical problem is one of two-dimensional geometry, the kind commonly taught in school. What it boils down to is finding the two-dimensional shape that can be repeated endlessly to cover a large flat area (for bees, an entire honeycomb; for mathematicians, an entire two-dimensional plane), for which the total length of all the cell perimeters is the least, with the result that the total area of the honeycomb walls is as small as possible.

Mathematicians established some facts with ease. For instance, there are only three kinds of regular polygons that can be fitted together side by side to cover a plane: equilateral triangles, squares, and regular hexagons. (A regular polygon is a straight-edged figure whose edges are all the same length and whose angles are all the same.) Any other regular polygon will leave gaps. Of the three space-filling regular polygons, squares give a smaller total perimeter than triangles and hexagons do even better than squares.

Regular hexagons (i.e., the ones with equal sides and angles all 120 degrees) give a smaller perimeter than non-regular ones. This has been known for many hundreds of years. But if you allow combinations of polygons of all kinds, or edges that are not straight lines, then things rapidly become a lot more complicated. For this general situation, relatively little was known until 1943, when a Hungarian mathematician named L. Fejes Toth used an ingenious argument to prove that the regular hexagon pattern does indeed give the smallest total perimeter for all patterns made up of any combination of straight-edged polygons. (See figure 5.2.)

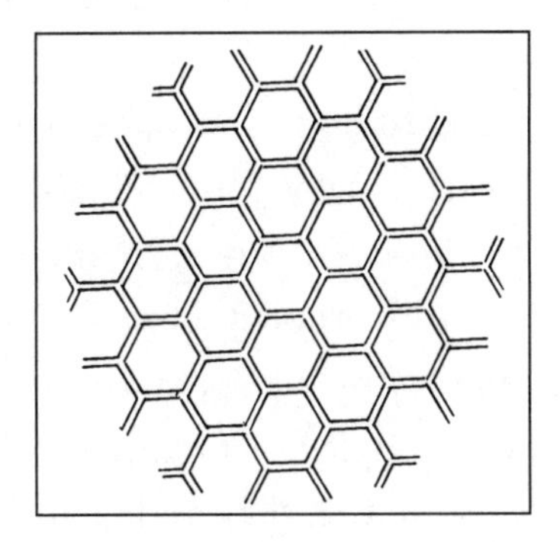

Figure 5.2. Mathematicians have proved that the cross-sectional shape that uses the minimum wax to store a given volume of honey is a repeating pattern of regular hexagons.

But what happens if the edges can be curved? Toth thought that the regular hexagon pattern would still be

more efficient than anything else, but he could not prove it.

In a single honeycomb cell, if a wall bulges out, you can store more honey in that cell for the same wall area than if the wall were straight. So, cell for cell, walls that bulge out are a more efficient way to store honey. But when all the cells are packed together, a wall that bulges out for one cell bulges in for the adjacent cell, resulting in less honey stored in that second cell. The question is, could there be an entire honeycomb of cells with bulging walls where the net increase in efficiency of the outward bulges outweighs the net decrease caused by the inward bulges? If there were such a pattern, the honeycomb conjecture would be false.

Intuitively, outward bulges would exactly balance out inward bulges, which is why Toth thought the hexagonal pattern would be the best. But as mathematicians who looked carefully at this problem observed, things are not quite as simple as they might seem. Nevertheless, that is exactly what Thomas Hales of Michigan proved in 1999: the bulges do cancel out. It took Hales nineteen pages of complicated mathematical argument to write out his proof. Mathematicians the world over were ecstatic when they learned of the new result. As for the honeybees, in their own way, they had known the theorem all along.

Now, if you struggled with high school mathematics you might marvel at how such a seemingly lowly creature as the honeybee can perform a mathematical feat that

took professional mathematicians so much effort. What mathematicians found really hard was proving that the honeycomb was the most efficient shape. All the bees have to do is construct the honeycomb. But of all the different architectures they could have used, Hales's theorem shows that the structure they use is the most efficient, so in fact the evolution of the honeybee incorporated a natural proof of that result.

Leaving aside the issue of the efficiency of the repeating-hexagon shape, however, the incredible precision with which the bees construct their honeycombs means they are natural geometers and engineers of the highest order.

As with the Tunisian desert ant and migrating birds and fish, hundreds of thousands of years of evolution have produced a creature whose natural instincts make it a perfect construction machine, including planning, computing, measuring, and implementation. To be sure, humans can do all of those things–indeed, we can do them all with far greater precision than the honeybee. But not by instinct. Rather, only with the explicit, conscious use of a fairly sophisticated degree of mathematics.

When it comes to construction, then, the humble honeybee seems to be a much more natural engineer than any human architect or builder. But constructing honeycombs is not the only feat in the bees' mental repertoire. Nature has also endowed them with an elegant and effective communication system and a mathematically sophisticated

means of judging distance, both of which they use in order to obtain food. Here is what they do.

Honeybees are social creatures. They live in large colonies and divide up the daily chores. While some bees stay at home and concentrate on constructing and maintaining the beehive and the honeycomb, others have the task of supplying the colony with food. They have evolved a highly efficient way to do this. Forager bees fly out in search of a good food source. When one is found, the forager bee flies back to the hive and tells the others where the food is. It does this by performing a ritual dance that indicates the precise direction and distance to the food.[12]

If the food source is fairly close to the nest, say, no more than 50m (165 feet) away, the returning forager bee performs what scientists call a "round dance," in which the bee simply flies around in a circle. This tells the others that a food source has been found, but gives no indication of where it is, leaving the forager to have to lead the others to it. If the food source is farther than 50m away, however, then in order for the forager not to have to make a long trip again it performs a more complicated dance, called the "waggle dance," that indicates both the precise direction and the distance to the food. (See figure 5.3.) Other bees can then locate the food source without the guidance of the forager.

12. This was first described by K. von Frisch in his book *The Dance Language and Orientation of Bees*, London: Oxford University Press, 1967.

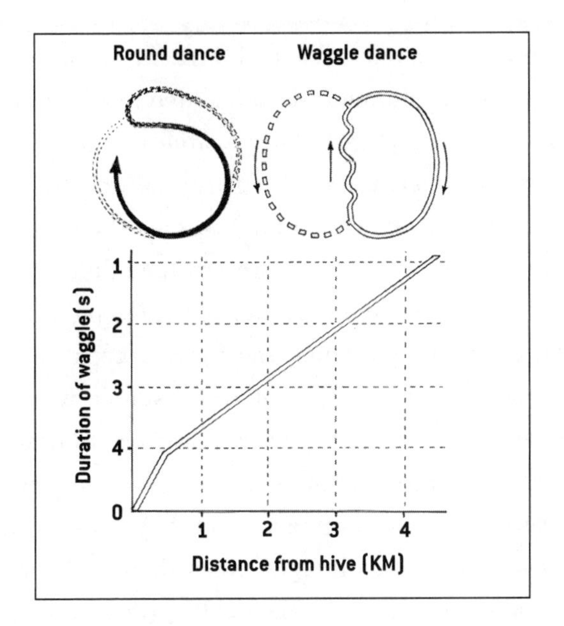

Figure 5.3: (Top) The round dance and the waggle dance of the foraging honeybee. The angle of the waggle component indicates the direction of the food source with respect to the sun's azimuth. The duration of the waggle indicates the distance to the source. The relationship between the duration of the waggle and the distance to the food is what mathematicians call "piecewise linear." The time of the waggle increases linearly at a rate of roughly 2 sec per km for distance to source up to just under 0.5km, thereafter increases linearly at a rate of about .7 sec per km. See graph at bottom.

We learned in the previous chapter that bees are able to determine orientation by detecting the polarization of sunlight. Thus, the waggle dance can tell them the direction to fly in order to reach the food. But how do they determine the distance? They certainly know this, because experiments in

which researchers move the food farther away after the initial foragers have found it have resulted in the subsequent bee collection party's having to search in vain when they reach the spot where the food had originally been.

K. von Frisch, who originally studied the bee dance in the 1960s, thought that bees determined distance by the energy they expended on the journey. He based his conclusion on the fact that, when the forager bees flew into a headwind, their estimation of the distance was greater than it would otherwise have been. But two studies carried out in the mid 1990s showed that this was not the case. Bees use vision to determine how far they have flown. They note how fast various images cross their retinas during flight, a phenomenon scientists call *optic flow.* Optic flow tells the bees their speed relative to the ground or to other landmarks in their visual field, and they can compute the distance by combining that with the time that has elapsed.

Of course, the speed with which the visual field moves by depends on how far away you are from the images you are seeing, as anyone knows who has noticed how slowly the ground seems to flow backward when viewed from an airplane at 30,000 feet, compared with how fast the walls fly past when a train passes through a tunnel. The farther away you are, the slower seems the motion. This is true for bees as well. Flying a short distance close to the ground generates the same distance information as flying a longer distance high above the ground.

In one study, led by H. E. Esch and J. E. Burns of

Notre Dame University,[13] the experimenters trained one group of bees to forage from the top of a 50m (165 feet) high building to a feeder placed on the roof of another tall building 230m (about 750 feet) away. The waggle dance indicated a distance about half the length signaled by another group of bees traveling the same distance at ground level from a hive on the ground to a food source at ground level.

Another study, carried out by Mandyam Srinivisan and his colleagues at the Australian National University,[14] used a different method to reach the same conclusion. They trained bees to fly several meters down a well-lit tunnel to forage at a feeder placed a measured distance along its length. On some occasions, the inside of the tunnel was marked with a pattern, either random markings or else rings perpendicular to the direction of flight. With such a configuration, when the feeder was removed, the bees kept searching for it exactly where they expected it to be, indicating that they knew how far they had flown along the tunnel. But when the tunnel walls were unmarked, or marked with lines the same thickness as the rings but running horizontal, straight along the tunnel in the direction of flight (and thus providing no visual sense of motion),

13. H. E. Esch and J. E. Burns, *Journal of Experimental Biology* 199 (1996). p. 155.

14. Mandyam Srinivisan, Shaowu Zhang, Monika Altwein, and Jürgen Tautz, "Honeybee Navigation: Nature and Calibration of the 'Odometer,' " *Science* 287 (2000), p. 851.

the bees were unsure of the distance they had flown, and kept flying, searching in vain for the missing food.

The bees tended to fly along the center of the tunnel. By varying the diameter of tunnels with random or perpendicular ring patterns, the bees were forced to fly at differing distances from the visual landmarks. The smaller the diameter of the tunnel, the less far the bees would travel before they expected to reach the food. This indicated that they interpreted motion past the closer landmarks as flying farther than motion (at the same actual speed) past landmarks farther away in tunnels of greater diameter—just as we experience when we compare passing through a tunnel in a train with flying high above the ground in an airplane. Clearly, energy consumption plays almost no role in bee navigation; their distance measurement is based almost entirely on optic flow.

By tabulating the results obtained by observing bees fly through randomly marked tunnels of differing diameters, the researchers were able to work out the mathematical equations (using trigonometry and basic calculus) that the bees must implicitly solve to determine distance from optic flow. By solving those equations, the researchers were able to conclude that, for a bee, flying a short distance through a randomly marked narrow tunnel is equivalent to flying a distance as much as thirty times farther over open ground. The fact that the conditions of the experiment had such a huge effect on the distance the bees flew shows the amazingly high degree to which they depended on mathematics (their innate, built-in trigonometry and calculus) to

make their normally highly accurate determination of distance, based on optic flow.

But now we have allowed ourselves to wander off course—something the honeybee generally does not. Our focus in this chapter is construction. There is no animal builder more famed than the beaver and his dam. But is that fame properly earned?

Who gets the credit for the beaver's dam?

Should a beaver get the credit for building a dam? To put it another way, has nature endowed this creature with the mathematical and engineering knowledge to place branches, sticks, twigs and dirt across a stream in such a way that it will block the flow of water?

We should be careful in ascribing mathematical powers to creatures that perform a particular activity. The issue is, where is the mathematics being done, and what is doing it? In the examples we've looked at so far, the animals have all been equipped with capacities that, when viewed in human terms, can only be described as "doing math." In those cases, nature, through the mechanism of natural selection, has produced an animal that is "hardwired" to perform a particular mathematical computation—or, if you prefer, the animal instinctively carries out a process that humans can perform only by using mathematics.

But there is another possibility: It could be said that the environment does the math. If you step off the top of a

tall building, you will fall to the ground. Would you say that in doing so you were solving Newton's equations for a body's motion under gravitational force? Surely not. In falling, you are not doing any math; you are simply obeying the physical laws that Newton expressed with mathematics. The math, insofar as there is any, is being done *on you* by the universe.

Figure 5.4. The beaver and his dam. Credit for the dam should probably go to the river as much as to the beaver.

This seems to be the case with the beaver (figure 5.4). As far as we can tell from observations, the only dam-building ability beavers are endowed with is an instinct to collect bits of trees, bushes, and dirt and pile them together across a stream. It's the steady force of the flowing water that forces the debris tightly together to create a dam. Likewise, it is the flow of the water that makes the dam settle into a compact, efficient shape that

looks as though it had been carefully designed. Of course, you could say, "What a clever beaver, figuring out how to make use of the water flow in that way." But there is absolutely no evidence that this is what is going on. Most likely we will never know exactly what the beaver thinks it is doing when it goes about its business of building dams, or whether it has any conscious thoughts at all. Given the parsimonious nature of evolution, however, since the beaver-plus-stream symbiosis provides an efficient dam-building system, it seems likely that the beaver is simply following an instinct to collect debris and put it in the stream, and the dam itself results from the pressure of the flowing water. And that means that beaver dam-building is more like "solving" Newton's laws of motion by jumping of a cliff than the animal doing anything that could reasonably be classified as "natural math."

Spinning a Web

Spider webs, for all that they look highly geometric, turn out to be more like beaver dams than bees' honeycombs. Certainly, it seems we should credit the spider with considerable innate engineering skill. But there appears to be no reason to assume that the elegant geometry of the final web is the result of some built-in spider mathematics. The overall shape of the web is the result of the spider performing some very simple steps. Mathematics can show

how those simple steps, when repeated, give rise to the web, but in this case too any mathematical credit should go to nature, which "programmed" the spider to perform those basic steps. Here is what is going on.

There are at least two thousand species of spider in the United States alone, but only a few build elaborate webs. The webs fall into four types: orb webs, sheet webs, funnel webs, and the irregular festoons built by house spiders. In all cases, it is the female that constructs webs. Let's focus on those intricate, beautiful, and geometrically precise orb webs constructed by the familiar, big, black-and-yellow garden spider. Figure 5.5 shows a picture of such a web.

Figure 5.5. The orb web of the garden spider. The elegant, geometric shape is the result of the spider repeating some very basic moves.

It takes a garden spider between one and three hours to build her web, a task she usually performs at night. Its purpose is to catch insects for food. On each of its hind legs the spider has a row of curved bristles that she uses to fling strands of silk over any insect that gets caught in the web. Once the prey has been rendered helpless, the spider literally sucks the life–that is, all the body liquids–from it. Occasionally, a larger creature such as a baby mouse will blunder into the web, in which case it may meet the same fate.

Although the spider has eight eyes, she builds her web almost entirely by touch. Beneath her abdomen are six fingerlike appendages called spinnerets, which she uses to produce and manipulate the silk to make the web. Each spinneret has several tiny outlets, which produce different kinds of silk in liquid form. In some parts of the web, the spider uses a single strand of silk. For the main structural strands, however, she produces multi-filament strands, much like rope. The silk solidifies as soon as it comes into contact with the air, to form a strand that is about five times stronger than a steel fiber of the same size, and yet may be stretched up to 30 percent of its original length without breaking. (Scientists have studied the chemical composition of spider silk–it is made up of chains of amino acids, primarily glycine and alanine–to try to copy it in the laboratory. The goal is to create a similar material for use in car seatbelts, parachute chords, and so on. This goal has not yet been achieved.)

To build its web, the garden spider must first find two vertical supports, between which the structure will be strung. It needs to choose those carefully, for the first part of the construction process requires a bit of help from nature. That tricky initial challenge is to construct the first strand, which connects the two main supports. Here the creature relies somewhat on luck, although a wise choice of where to start can increase its chance of success enormously.

The spider climbs the first of the two supports, attaches one end of a multi-filament silk strand, and then spins the strand downward, all the while hanging freely from the lower end. Then, when the spider feels the line is long enough, it stops spinning and simply hangs there, waiting for a gentle gust of wind to swing it across to the other support. As soon as that wind comes, the spider grabs ahold of the second support and attaches the free end of the initial thread. The silk is so fine and light that it does not require more than the most gentle air movement to achieve this step. What requires skill on the part of the spider is, first, identifying the two supports, and second, judging how long to spin the strand so that it will reach the second anchor point.

Once the first thread has been put in place, the spider can use it as a bridge to cross from one side another. The next step is to run a second thread down from the midpoint of the first to form a Y, anchoring the vertical strand either to the ground or to some other suitable support.

With the Y in place, the spider then constructs more

radial arms from the center point outward. This is a fairly intricate process, because at the same time the spider constructs an outer-frame thread, which it affixes to various anchor points.

Next, the spider positions itself at the middle point of the star-shape it has just created, and starts to spiral out, spinning a long continuous thread, which it joins to each arm of the star as it crosses. This first spiral is a temporary structure, to hold the web in place during the more elaborate phase of the construction that comes next.[15]

With the temporary spiral in place, the spider then starts at a point on the outer edge of the web and begins to spin the inward running capture spiral, a denser structure made of the sticky silk that will trap its prey. During this part of the operation, the spider removes the threads that make up the temporary spiral. In constructing the capture spiral, the spider appears to be guided primarily by the need to have sufficient density of strands. To achieve this, it tries to keep the distance between successive turns equal, producing a structure that mathematicians call an arithmetical spiral.[16] The main goal, however, is a web of sufficient

15. In the temporary spiral, the distances between successive turns increase by a constant growth factor. This gives a spiral that grows outward rapidly, having few complete turns. Mathematicians call such a figure a logarithmic spiral. It is characterized by having the same angle of growth at each point.

16. In an arithmetic spiral, a line drawn outward from the center crosses each successive turn at the same distance, and the tangent angles increase at a constant rate, approaching 90 degrees as the number of turns increases.

density, not an elegant geometry, and to accomplish this in a web whose outer frame is unlikely to be perfectly symmetrical, the spider has to improvise, occasionally even doubling back to create U-turns in the spiral.

When it has finished its labors, the spider heads back into the center and waits for dinner to arrive.

What we see as geometric elegance is, then, the result of three basic construction steps: a star; a temporary, constant-angle spiral; and a much denser, constant-distance capture spiral. It is undoubtedly an impressive feat of engineering, in design, execution, and the strength of the materials used. To carry out the construction, the spider has to be good at judging distances. But the mathematical aspects of the web do not have to be calculated; they are automatic consequences of the simple steps the spider follows.

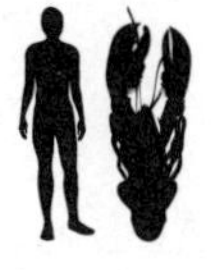

6.

Natural Artists: The Animals (and Plants) That Create Beautiful Patterns

Except for cases such as the beaver or the spider, where the environment "does the math," all of the examples we have encountered so far have one thing in common: the creature performs some activity or exhibits some behavior that we humans can perform—or describe—only by using mathematics. I explained this by saying that, through the mechanism of evolution by natural selection, nature has equipped the creature with the capacity to "do the (natural) math." The examples I shall introduce in this chapter and the next are different. In these cases, nature has arranged things so that the animal or the plant follows some specific mathematical rules as it grows and develops.

How does the leopard get its spots?

Figure 6.1. The tiger and the leopard. Their coat patterns are the result of mathematical rules followed by pigmentation chemicals in the skin.

Have you ever wondered how the leopard gets its spots, or the tiger its stripes? In the late 1980s, James Murray of the University of Oxford did just that. Being a mathematician with a considerable knowledge of biology, he was able to find an answer. Animal coat patterns provide an example of another way that nature can "do math" in a way different from the instances we have seen so far.

Murray knew the familiar explanations of why different

animals have different coat patterns—a combination of camouflage in the animal's natural habitat, maintaining an appearance that wards off certain predators, and looking attractive to members of the opposite sex. He knew too that any coloration of an animal's coat is caused by a chemical called melanin, which is produced by cells just beneath the surface of the skin. (It's the same chemical that makes fair-skinned people develop a suntan.) But what mechanism did nature use to put the melanin in the right places to produce the skin coloration typical for that animal?

A possible answer is that the animal's DNA encodes all the information necessary to generate the coloration—specific instructions of where to put a spot, where a stripe, and so forth. But there is another possibility, one that, if it works, would be much more efficient. Suppose, said Murray, there are geometric rules that govern animal coat patterns, a bit like the rules of the ordinary geometry of triangles, circles, tetrahedra, and the like that the Greek geometer Euclid (ca. 300 B.C.) wrote down. Then all the animal's DNA would need to encode would be some instructions about which rules to apply and when. The mathematics would then generate the actual coat pattern.

On the face of it, this might seem unlikely. After all, aren't the kinds of shapes that mathematical formulas can describe very regular? Not at all like animal coat patterns. In fact, that's not the case. Mathematicians can write down equations that describe the formation of animal coat patterns, and many other aspects of living things. What they

cannot do in general is solve them—at least, not with paper and pencil. But with a powerful computer, they can. And that is what Murray did for animal coat patterns.

His first step was write down equations that described the chemical processes that cause coloration in an animal coat.[17] The second step was to write a computer program to solve those equations. The third step was to use computer graphics to turn those solutions into pictures.

During the early stages of its development, the embryo of a leopard or a tiger, for example, has no skin pattern. But its skin contains chemicals that, although they do not themselves color the skin, do react together to produce the color-creating melanins. The crucial chemical reactions take place during the animal's early development. For most animals, born with their skin pattern intact, the reactions take place within the womb. For a few animals, born without their skin pattern, the reactions occur shortly after birth. (This happens with the dalmatian, whose spots appear when it is a few weeks old.)

The animal's DNA determines which melanin-producing chemicals occur in the skin, and what their relative concentrations are, but not where they are located. The initial distribution of these chemicals is random. The only other color-producing information the DNA encodes are two time triggers that instruct the growing embryo when to activate the chemicals and when to bring the reactions to a stop.

17. Murray's equations involve calculus. Mathematicians call them partial differential equations.

The surprising discovery Murray made using his computer simulations was that this simple mechanism is all that is necessary to generate all the different animal coat patterns we see in nature. The main factor that differentiates spots from stripes, for instance, is the timing of the chemical reactions within the skin.

In fact, it's not the timing itself that makes the difference, rather it is the size and overall shape of the embryo during the active phase. Here's what the mathematical equations predict: Very small or very large skin regions lead to no pattern at all. In the intermediate range, long, thin regions lead to stripes perpendicular to the length of the region, and squarish regions of roughly the same overall area gave rise to spots whose exact pattern depends on the region's dimensions. Figure 6.2 shows some of the skin patterns Murray obtained on his computer.

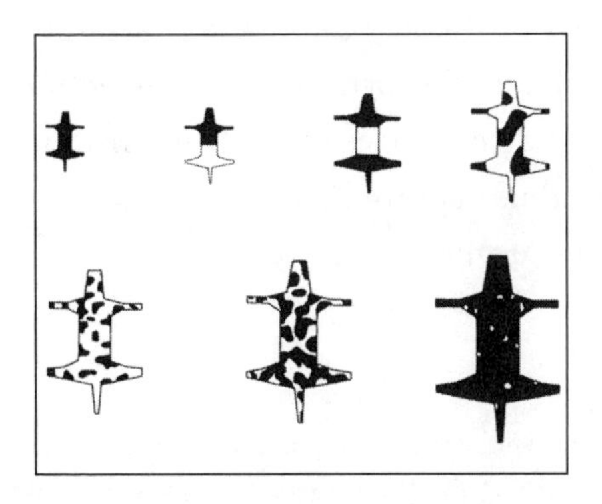

Figure 6.2. By varying two numerical parameters in his computer model, James Murray obtained all the animal coat patterns found in nature, suggesting that the variety of animal coat patterns is probably the result of mathematical rules.

For example, there is a four-week period early in the year-long gestation of a zebra during which the embryo is long and pencil-like. Murray's mathematics shows that, because the reactions take place during this period, the resulting pattern is stripes. Leopard embryos, however, are fairly chubby when the reaction occurs, and so the equations give rise to spots. (Apart from the tail, that is. The tail is long and pencil-like throughout development, which explains why the tail of the leopard is always striped.)[18]

Although the mathematics required to explain the process is amazingly simple, the mechanism is very sophisticated—Murray's equations involve methods of calculus—but neither the mother nor her babies are doing that math. Rather, nature is exploiting the math in order to provide an extremely efficient mechanism to generate coat patterns. (Moreover, since the key parameters are the two timing triggers, nature, in the form of natural selection, can easily change the pattern in case the environment should change, and thus require a change in skin coloration for the species to survive.)

18. Murray's mechanism also provides an answer to a puzzling question: Why is it that several kinds of animals have spotted bodies and striped tails, but none have striped bodies and spotted tails? There seems to be no evolutionary reason for this curious fact. Given Murray's mathematics, the explanation is simple. It's a direct consequence of the fact that many animal embryos have chubby bodies and skinny tails, but no animal embryo has a long skinny body and a chubby tail.

The Nautilus and the Peregrine Falcon

Another example of how natural patterns are a result of hidden mathematical laws is provided by the nautilus shell, that beautiful spiral-shaped horn that we press to our ears when we find one on the beach, to hear the sound of the ocean—an odd behavior given that the waves are just a few feet away. (See figure 6.3.)

Figure 6.3. The chambered nautilus. The shell of the creature is shown here sliced in half.

The chambered nautilus is the only surviving descendant of the nautiloids, 450 million years ago the largest predators in the seas. It lives in the tropical waters of the Indo-Pacific Ocean. Its familiar smooth, coiled shell, shown in figure 6.3, can grow up to eleven inches in diameter. It is

separated into a series of progressively larger compartments, each lined with mother-of-pearl, the outer (and most recent) of which is inhabited by the animal. The walls dividing the chambers are pierced by a tube connected to the nautilus, and the nautilus regulates its buoyancy by passing gas and liquid into and out of the chambers through the tube's walls. Nautiluses spend most of their time at depths of 600 to 800 feet, but rise to 200 feet at night to feed.

The nautilus shell's spiral shape is, like the temporary spiral of the garden spider's web, what mathematicians call a logarithmic spiral. There are several equivalent ways to describe it mathematically. One way is to say that it is an equi-angular spiral: the angle of the curve remains constant throughout the length of the spiral. Another description is that it is self-similar: if you take one complete revolution of the spiral and enlarge it, you will find it fits exactly over all larger revolutions.

The self-similar property of the spiral is the reason why the nautilus shell is that shape. As the nautilus grows, it must enlarge its living quarters. Since the creature does not change shape, but simply grows larger, the most efficient way to do this is for its shell to grow in the self-similar form of a logarithmic spiral.

Figure 6.4. The peregrine falcon follows a logarithmic spiral as it races down to catch its prey.

Another place where the logarithmic spiral crops up in nature is that it is the path followed by a peregrine falcon when it swoops down on its prey. The obvious question is, why doesn't the falcon simply dive right for the target? The answer is that the falcon must keep the prey in its sight all the time. But there is a problem. Although the falcon's eyes are razor sharp, they are fixed on either side of its head. So what the creature does is swivel its head to one side, by an angle of about 40 degrees, and fix the prey in one eye. Keeping its head fixed at that 40-degree angle, the falcon then dives in a way that keep the prey in view in that one eye. The fixed angle of the head results in the bird following an equi-angular spiral path that converges on the prey. A natural geometer–not unlike plants, in fact, as we'll discover next.

THE NUMERICAL PATTERNS THAT PLANTS WEAVE

Can plants do math? In the usual sense, no, of course not. They don't have a brain. But as we have seen, living things sometimes solve mathematical problems or generate mathematical patterns just by the way they grow or behave. Leopards, tigers, and the chambered nautilus are all examples of nature's own calculators. In a similar way, so too are many flowers and plants.

Our plant story begins not in the garden but with an arithmetic problem in a thirteenth-century arithmetic textbook. In 1202, the great Italian mathematician Leonardo of Pisa (ca. 1170-1250, whom historians subsequently called Fibonacci) completed an arithmetic textbook called *Liber abaci* (The Book of Calculation). One of the problems in that book read as follows:

> A certain man had one pair of rabbits together in a certain enclosed place, and one wishes to know how many are created from the pair in one year when it is the nature of them in a single month to

bear another pair, and in the second month those born to bear also.[19]

As in most mathematics problems, you are supposed to ignore such realistic happenings as death, escape, or impotence. Fibonacci stated the problem purely as an exercise in mathematics to help readers of his book.

After some thought, you see that the number of pairs of rabbits in Fibonacci's garden in each month is given by the numbers in the sequence 1, 2, 3, 5, 8, 13, 21, 34, 55, 89, 144, and so on. This sequence of numbers is called the Fibonacci sequence. The general rule that produces it is that each number after the second one is equal to the sum of the two previous numbers. (So 1 + 2 = 3, 2 + 3 = 5, 3 + 5 = 8, and so on.) This corresponds to the fact that each month, the new rabbit births consists of one pair to each of the newly adult pairs, plus one pair for each of the earlier adult pairs. Once you have the sequence, you can solve Leonardo's problem by simply reading off the twelfth number in the sequence: after one year there will be 233 pairs. (Given the way the problem is stated, there is some uncertainty as to whether you should count to 12 or 13; Leonardo himself counted to 13 and gave the answer as 377 pairs.)

19. Many books paraphrase this problem. The text I give here is a direct translation from the original Latin, taken from L. E. Sigler's complete annotated translation *Fibonacci's Liber abaci*, Springer Verlag (2002), p. 404.

The seventeenth-century astronomer Johannes Kepler (1571-1630) seems to have been one of the first people to notice that the Fibonacci numbers seem to occur in nature. They do so in various surprising ways. For instance, if you count the number of petals in different flowers you will find that the answer is often a Fibonacci number. This happens much more frequently than you would get by chance. For instance, an iris has 3 petals; primroses, buttercups, wild roses, larkspur, and columbine all have 5 petals each; a delphinium has 8; ragwort, corn marigold, and cineria each have 13; asters, black-eyed Susan, and chicory have 21; daisies have 13, 21, or 34; a plantain and a pyrethrum each have 34; and Michaelmas daisies have 55 or 89 petals—all Fibonacci numbers.

For another example from the botanical world, if you look at a sunflower you will see a beautiful pattern of two spirals, one running clockwise, the other counterclockwise. Count those spirals and for most sunflowers you will find that there are 21 or 34 running clockwise and 34 or 55 counterclockwise, respectively—all Fibonacci numbers. Less common are sunflowers with 55 and 89, with 89 and 144, and even 144 and 233 in one confirmed case. Other flowers exhibit the same phenomenon; the herb echinacea is a good example. Similarly, pine cones have 5 clockwise spirals and 8 counterclockwise spirals. A pineapple has 5, 8, 13, and 21 spirals of increasing steepness. Each scale on the pineapple is part of 3 different spirals. (See figure 6.5.)

Figure 6.5. The seeds in the heads of sunflowers and various other flowers exhibit two spiral patters running in opposite directions. The number of spirals going each way is always a Fibonacci number. A similar spiral pattern can also be found in pine cones, where the numbers of spirals are also Fibonacci numbers.

One further example concerns the way the leaves are located on the stems of trees and plants. If you take a look, you will see that, in many cases, as you progress up along a stem, the leaves are located on a spiral path that winds around the stem. The spiral pattern is sufficiently regular that it leads to a numerical parameter characteristic for the species, called its divergence. Start at one leaf and let p be the number of complete turns of the spiral before you find a second leaf directly above the first, and let q be the number of leaves you encounter going from that first one to the last in the process (excluding the first one itself). The quotient p/q is called the divergence of the plant. This is illustrated in figure 6.6.

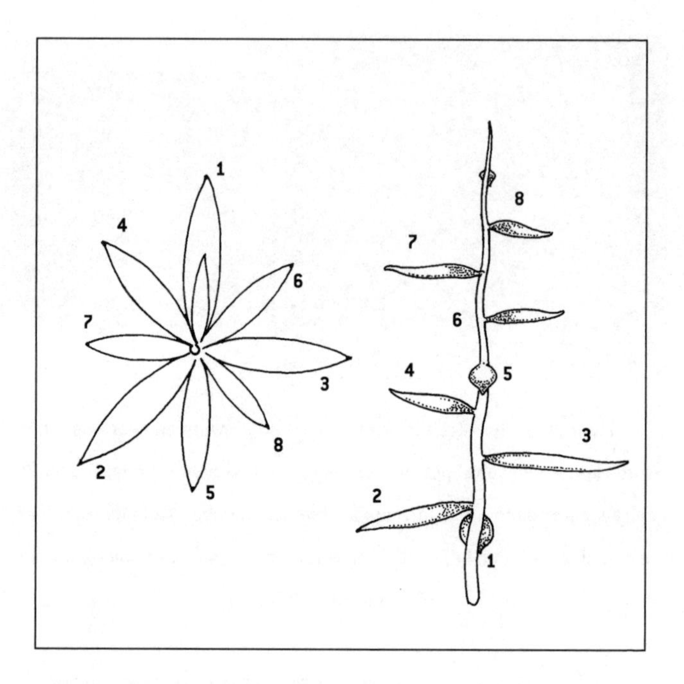

Figure 6.6. The leaves on a plant stem wind around the stem in a way that obeys precise mathematical laws involving the Fibonacci numbers.

If you calculate the divergence for different species of plants, you find that both the numerator and the denominator tend to be Fibonacci numbers. In particular, $1/2$, $1/3$, $2/5$, $3/8$, $5/13$, and $8/21$ are all common divergence ratios. For instance, elm, linden, lime, and some common grasses have a divergence of $1/2$, beech, hazel, blackberry, sedges and some grasses have $1/3$, oak, cherry, apple, holly, plum, and common groundsel have a divergence of $2/5$, poplar, rose, pear, and willow have $3/8$, and almonds, pussy willow, and leeks come in at $5/13$.

None of the examples I have given are numerological coincidences. As I'll explain below, they are consequences of the way plants grow. (For example, the leaves on a plant stem should be situated so that each has a maximum opportunity of receiving sunlight, without being obscured by other leaves.) The Fibonacci sequence is one of a number of very simple mathematical models of growth processes that happens to fit a large variety of real-life growth processes.

In addition to its connections with the natural world, the Fibonacci sequence has a number of curious mathematical properties. Perhaps the most amazing is that it is closely connected to the famous "golden ratio" number $\Phi = 1.61803\ldots$, the "perfect proportion" ratio said to be much beloved by the ancient Greeks. (Φ is the capital Greek letter *phi,* pronounced "fye." The use of Φ to denote this number is fairly recent.)

According to an oft-repeated story, the ancient Greeks believed that the most pleasing rectangle is one where the ratio, x, between the two sides is obtained like this. Take a straight line AB and divide it into two by a point P so that the ratio AP:PB is x:1, as shown below (where we choose units so that PB has length 1, to simplify the mathematics).

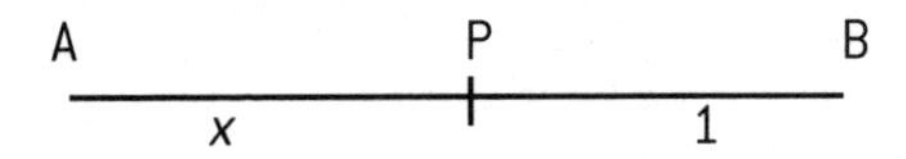

Then for the rectangle with long side equal to AP and short side equal to PB to be the most pleasing to the eye, the ratio (x) of the longer segment AP to the shorter one PB should be exactly the same as the ratio of the entire line AB to the longer segment AP. As a slogan: *Whole is to longer as longer is to shorter.* In symbols:

$$\frac{AB}{AP} = \frac{AP}{PB}$$

It doesn't matter what units you use (i.e., what the actual length of the line AB is), 1 foot, 1 meter, 1 shoelace length, etc. For the perfect rectangle, it is the ratio between width and height that counts, what modern designers call the aspect ratio, not the actual lengths themselves. That's why we can take the length of PB to be 1.

To find the golden ratio, we now have to do a bit of algebra. Since the length of AP is x and the length of PB is 1, the length of AB will be $x + 1$.This means we can rewrite the above geometric identity as the equation

$$\frac{x+1}{x} = \frac{x}{1}$$

This can be rearranged by cross multiplying to give

$$1(x+1) = x\,x$$

that is

$$x + 1 = x^2$$

We can rearrange this to give the quadratic equation

$$x^2 - x - 1 = 0$$

If you remember back to your high school algebra class, quadratic equations have two solutions, and there is a formula to give you those solutions. When you apply this formula to the above equation, you get the two answers:

$$x = \frac{1 + \sqrt{5}}{2} \text{ and } x = \frac{1 - \sqrt{5}}{2}$$

These don't work out exactly. Using a calculator, to three decimal places, the answers are 1.618 and -0.618, respectively. The golden ratio, **Φ**, is the first of these two solutions, the positive one.

You start to suspect that there's more to **Φ** than meets the eye when you ask what happened to the negative solution to the quadratic equation, -0.618 . . . It too goes on forever as a decimal. Apart from the minus sign it looks just the same as the first solution (**Φ**) but with the initial 1 missing, and indeed that is the case. The negative solution is in fact equal to $-1/\Phi$. That certainly doesn't usually happen with quadratic equations. Maybe the Greeks were on to something when they thought this particular number worthy of study.

Having found their golden ratio, the story continues, the Greeks incorporated it into their architecture, ensuring that wherever they went in their glorious cities, their eyes would be met with so-called divine (or golden) rectangles.

This may be true, but modern historians question the claim. Certainly, the oft repeated assertion that the Parthenon in Athens is based on the golden ratio is not supported by actual measurements.

In fact, the entire story about the Greeks and the golden ratio seems to be without foundation. The one thing we know for sure is that Euclid, in his famous textbook *Elements*, written around 300 B.C., showed how to calculate its value. But his interest seemed more that of mathematics than architecture, for he gave it the decidedly unromantic name "extreme and mean ratio." The term "divine proportion" first appeared with the publication of the three-volume work by that name by the fifteenth-century mathematician Luca Pacioli. Calling **Φ** "golden" is even more recent: in 1835, in a book written by the mathematician Martin Ohm (whose physicist brother discovered Ohm's law).

Whether or not the ancient Greeks felt that the golden ratio was the most perfect proportion for a rectangle, many modern humans do not. Numerous tests have failed to show up any one rectangle that most observers prefer, and preferences are easily influenced by other factors. Claims that architects have based their designs on the golden ratio also fail to stand up to analysis, although the French architect Le Corbusier did at one stage become quite enthusiastic about its use.

It is true that several artists have flirted with **Φ**, but again you have to be careful to distinguish fact from fiction.

The oft-repeated claims that Leonardo da Vinci believed that the golden ratio is the ratio of the height to the width of a "perfect" human face and that he used Φ in his famous illustration known as *Vitruvian Man* seem to be without foundation. So too is the equally common claim that Botticelli used Φ to proportion Venus in his famous painting *The Birth of Venus.* Painters who definitely did make use of Φ include Paul Sérusier and Gino Severini in the nineteenth century and Juan Gris and Salvador Dalí in the twentieth, but all four seem to have been experimenting with Φ for its own sake rather than for some intrinsic aesthetic reason.

Unlike all the false claims made about the golden ratio in aesthestics, art, and architecture, the golden ratio definitely does play a fundamental role in the way flowers and plants grow.

Ever one for efficiency, nature seems to use the same pattern to place seeds on a seedhead, to arrange petals around the edge of a flower, and to position leaves around a stem. What is more, all of these maintain their efficiency as the plant continues to grow. How exactly does this happen?

Plants grow from a single tiny group of cells right at the tip of any growing plant, called the meristem. There is a separate meristem at the end of each branch or twig, and this is where new cells are formed. Once formed, they grow in size, but new cells are only formed at such growing points. Cells earlier down the stem expand and so the

growing point rises. In order to achieve the best possible packing and to receive maximum sunlight, these cells grow in a spiral fashion, as if the stem turns by an angle before a new cell appears. These cells may then become a new branch, or may become petals and stamens on a flower.

Amazingly, a single fixed angle of rotation can produce the optimal design, no matter how big the plant grows. In the case of leaves, the angle will ensure that each leaf will least obscure the leaves below and be least obscured by any new leaves above it. Similarly, once a seed is positioned on a seedhead, the seed continues out in a straight line pushed out by other new seeds, but retaining the original angle on the seedhead. No matter how large the seedhead, the seeds will always be packed uniformly on it.

As early as the eighteenth century, mathematicians suspected that the single angle of rotation that can make all of this happen in the most efficient way is the golden ratio (measured in number of turns per leaf, etc.). However, it took a long time to put together all the pieces of the puzzle, the final step coming in 1993 with some experimental work of two French scientists, Stéphane Douady and Yves Couder.

Today, we know why Φ plays such a crucial role in plant growth. The scientific part of the explanation is that it is the ratio that gives the optimal solution to the growth equations. The mathematical explanation behind the

science is that of all irrational numbers, **Φ** is in a very precise, technical sense the farthest away from being representable as a fraction.[20]

This explains why there are so many occurrences of the Fibonacci sequence in flowers and plants. The key is the close connection between the Fibonacci sequence and the golden ratio.

What exactly is that connection? As you proceed along the Fibonacci sequence, the ratios of the successive terms (i.e., $2/1 = 2$, $3/2 = 1.5$, $5/3 = 1.666$, $8/5 = 1.6$, $13/8 = 1.625$, $21/13 = 1.615$, $34/21 = 1.619$, $55/34 = 1.618$, etc.) get closer and closer to the golden ratio. From $55/34$ onward, the ratio gives the golden ratio to the first three decimal places of accuracy. Another way to express the same result is that the *n*th Fibonacci number is approximately equal to a fixed multiple of the *n*th power of the golden ratio. This gives a way to calculate the *n*th Fibonacci number without generating the entire sequence of preceding Fibonacci numbers: Take the golden ratio, raise it to the power *n*, divide by the square root of 5, and round off the result to the nearest whole number. The answer you get will be the *n*th Fibonacci number.

Thus, the reason you find the Fibonacci numbers everywhere you look in the plant world is that the Fibonacci

20. For those readers who are familiar with continued fractions, the continued fraction expansion for the golden ratio is [1;1,1,1, . . .]. That infinite sequence of 1s can be interpreted as meaning that **Φ** is the real number least like a fraction. This is the appropriate way to measure the degree of irrationality of irrational numbers in order to understand plant growth.

sequence is the sequence of whole numbers that most closely grows according to the golden ratio. Since the number of petals, spirals, or stamens in any plant or flower has to be a whole number, nature "rounds off" to the nearest whole number. In short: The Fibonacci numbers arise throughout the botanical world because the golden ratio is the rate of plant growth. Yet again, in its harmonic ordering nature shows us that it is a mathematician.

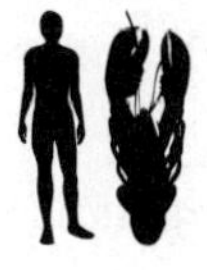

7.

It's Just a Step to the Right: The Math of Motion

A basketball player running at full speed suddenly stops, pivots on one leg, takes two steps in another direction, then launches himself high into the air to score a basket. A fish, motionless in the water one moment, catches a sudden movement in the corner of its eye and, with a barely perceptible flick of its tail, darts off rapidly into the safety of the reeds. A cat leaps elegantly from the floor onto the sideboard, a distance several times its own height, and lands softly and silently amid the glassware, causing not a single glass to topple and break. A hummingbird hovers at the mouth of a flower, its very motionlessness the result of a flapping of its wings too fast for the human eye to detect as anything other than a blur.

Our world is full of motion. Evolution has equipped most living creatures with a way to move in order to search for food, to seek out a mate, or to escape from danger. People and ostriches walk and run on two legs, horses and dogs on four, cockroaches on six, spiders on eight; snakes slither; fish propel themselves by pushing the water sideways with their tail; birds fly by flapping their wings to create lift and forward thrust. The tasty white meat from the shrimp plate that we dip in tomato sauce is a single muscle—some 40 percent of the total weight of the creature it belongs to—designed by nature for just one purpose: the creation of massive acceleration to propel the creature out of danger with a sudden explosive thrust. Similarly, the force generated by a squid when it ejects a jet of water at high speed to propel itself away from a sudden threat is powerful enough to impress a NASA rocket engineer.

How do the creatures that inhabit the land, the sea, and the air move? Recent research[21] has discovered that, despite the seemingly endless variety of different kinds of locomotion, all living creatures use a very similar process

21. A good survey is provided in the article "How Animals Move: An Integrative View," by Michael Dickinson, et al., *Science* 288, 7 April 2000, pp. 100-106, on which much of this chapter is based.

to generate motion. And when they move, they all make use of sophisticated (but, of course, implicit and hardwired) mathematics.

Locomotion brings us to a third way that mathematics arises naturally in the living world. In chapters 1, 2, 4, and 5 we considered examples where individual creatures were hardwired to perform certain computations in the course of their normal lives. Then, in chapter 6, we saw how the growth of an animal or plant can follow precise mathematical laws. In this chapter and the next, we will see how mathematics is built in to the *mechanical structure* of various animals. We begin with animal locomotion; then, in chapter 8, we look at the mathematics of vision. Both locomotion and vision involve some sophisticated mathematics, and as we shall see, nature has equipped practically all creatures (including humans) with very efficient mechanical "computers" to solve precisely the math problems necessary to get around and to see where they are going.

By land, sea, and air

You can get some idea of the difficulty of the mathematics of locomotion from the fact that after fifty years of well-funded research into the construction of computer-controlled machines, no one has yet been able to build a robot that can walk well on two legs. In fact, the best four-

or six-legged robots do not perform anything like as well as the average dog or dung beetle. Only the invention of the wheel thousands of years ago has enabled man to build efficient transportation machines. When it comes to building machines that imitate the ways nature solved the locomotion problem, we're still in kindergarten.

Yet all motion comes down to just two physical principles, identified by Isaac Newton some three hundred and fifty years ago. One is that motion results from the application of a force: Force = mass × acceleration. The other is that every force produces an equal and opposite reaction. The great variety of locomotive strategies that we see around us comes not from different principles of motion but from nature's boundless ingenuity in finding ways to apply Newton's two physical laws–ingenuity that required equipping various creatures with some highly sophisticated (built-in) mathematics.

Research carried out over the past five years or so has shown that the mathematics for motion is not all located in the creature's brain. Nature has provided its creatures with skeletons, muscular systems, and nervous systems that help the brain perform the math required for locomotion.

In fact, as we shall see shortly, the "math" a cockroach has to do to move is far more complicated and difficult that most of the math problems we typically solve on our calculator. Indeed, the cockroach is a particularly dramatic example of nature's mathematical prowess. The mathematical principles involved in cockroach locomotion

are very similar to those used to design and control the latest high-performance jet fighters.

In all creatures that move, locomotion starts with muscles. These are organs that are capable of repeated contractions. Those basic contractions have to be converted into a locomotive force. In many creatures, including *Homo sapiens* and all other mammals, this conversion is done by a system of levers, springs, and connecting rods—more precisely, bones, cartilage, tendons, and ligaments—that together with the muscles themselves make up what is known as the muscular-skeletal system. That conversion system, whatever form it takes, is what turns repeated contractions and relaxations of the creature's muscles into its locomotion.

Such conversations can require some sophisticated engineering. For comparison, consider the automobile. The "muscles" of a modern car are the combustion chambers. The repeated out-and-in motion of the pistons in each chamber provides the basic motive forces that drive the automobile. But it takes a fairly complex arrangement of rods, levers, and wheels (including those in the clutch and the gearbox) to convert that out-in motion of the pistons into forward motion of the car, and still more complex machinery (the acceleration, braking, and steering mechanisms) to ensure that the car goes in the direction we want, at the speed we want, when we want it to, and moreover that it does so smoothly, in a way that does not harm the occupants.

Designing a modern automobile requires a lot of mathematics. That mathematics is not simply "lost" or "forgotten" once the automobile rolls out of the factory and into the showroom. Rather, when the car is driven, the entire structure is continually "doing the math" required to drive it forward. We could, if we wished, view the entire drive train–all the rods, belts, cogs, wheels, and levers–as a computer, repeatedly performing the same calculations. We generally don't think of it that way, of course, because typically we don't regard a machine as a computer if it performs just one particular computation, or just a small collection of computations, time after time. Rather, we think of a computer as a device that we can program to carry out many different computations.

Nature has been no less ingenious than human engineers in designing mechanisms to convert muscle contractions into purposeful, orderly, directional locomotion. Those mechanisms often involve mathematics so sophisticated that, as we have already noted, human engineers have not yet succeeded in building a four-legged robot that walks as well as a dog, or a two-legged robot that can perform much better than a young toddler taking its first faltering steps.[22]

In the early days of robot locomotion, engineers

22. In this account, I shall largely ignore the role played by the animal's nervous and cardiovascular systems. In practice, these are also closely bound up with the action of the muscular-skeletal system, making the entire locomotive process even more complex than I describe.

adopted the approach they believed at the time that animals used: namely, the brain is the central control unit that coordinates the entire process, sending signals that direct the various muscles to act, and so forth. In recent years, however, scientists have discovered that nature has been much more efficient than that. The muscular-skeletal system of a mammal or an insect, for instance, is designed by natural selection so as to distribute the required locomotion computations over the entire structure, leaving the brain to focus on the more general issues, such as, where does the creature want to go and how fast? As a result of these discoveries, engineers now try to construct robots in a similar way, embedding much of the mathematics into the mechanical structure of the robot, leaving the controlling computer to handle the overall issues of the motion. (Many of the recent discoveries about animal locomotion have been made by surgically attaching tiny sensors to various muscles and joints in animals and birds, which send signals to computers to convey information that helps describe how the creature actually moves.)

In the case of the cockroach, for instance, its six legs actually work against each other for much of the time, pushing inward toward its body as well as providing an overall forward thrust. (See figure 7.1.) This gives it stability, making it resistant to sliding sideways down a slope or being blown over by a sudden gust of wind. It also enables the insect to change direction rapidly in order to avoid danger. A capacity for rapid change of direction is

likewise important for jet fighter aircraft, and aeronautic engineers achieve such maneuverability in a mathematically similar fashion by designing modern jet fighters to be intrinsically aerodynamically unstable, kept aloft and on course only by the generation of often mutually competing forces, controlled in real time by fast onboard computers.

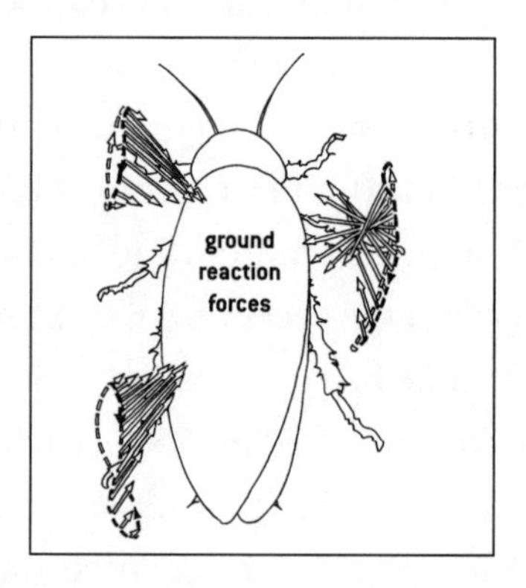

Figure 7.1. The six legs of a cockroach work against each other, pushing inward toward its body as well as providing an overall forward thrust.

The movement of the cockroach's six legs has to be very carefully coordinated to ensure that the periodic movement of the legs results in a fairly constant forward motion. Mathematically speaking, this requires the solution, in real time, of a complicated system of differential equations. The solutions to those equations represent instructions to the muscles controlling each leg to move the leg at a specific instant, what force to exert, and the length of time of the leg motion. For a human mathematician,

solving these equations would be a major challenge, even with the help of a powerful computer. For the cockroach the math presents no difficulty at all, of course, since it has evolved to be an automatic solver of precisely those equations.

But the math of motion doesn't end with the movement of the legs. For one thing, there is the question of what effect the motion has on the rest of the creature's body. In particular, creatures that walk upright on two or four vertical legs, such as humans, chimpanzees, dogs, or horses, have to handle the significant stresses that are repeatedly put onto the bones and joints of each leg when the creature's weight comes down to bear fully on that leg during locomotion. Those forces can be as much as 30 percent of the stress that would cause the bone to crack or break, which is far closer to the breaking point than civil engineers allow in buildings and bridges or mechanical engineers permit in machinery.

To ensure that locomotion does not result in broken bones, animals are constructed so that, when they go faster, they automatically change their gait so as to reduce the impact forces. Horses, for example, have four distinct ways of moving: walk, trot, canter, and gallop. So too do humans: walk, jog, run, and sprint. Monitoring and controlling such a range of different kinds of locomotion is a complicated task that requires a large amount of (built-in) mathematics. The locomotion control system is made more complicated by the necessity for balance control,

particularly for two-legged creatures, which are inherently unstable. This is achieved in large part by means of a complex system of sensors and feedback mechanisms. The sensors constantly monitor all aspects of the animal's position and motion and send signals back to the muscles to adjust the motion accordingly.

The difference between walking and (any kind of) running is particularly significant. In walking, the leg acts much like a rigid pole over which the animal vaults its weight. In running, the leg acts more like a sprung pogo stick, that compresses and stores up potential energy as the animal's weight bears down on the leg, and then converts that stored energy back into kinetic energy as the spring expands, propelling the animal's weight upward. (See figure 7.2.) Studies have shown that four-legged animals coordinate their legs in pairs, with each pair acting in one of these two ways, as rigid vaulting pole or sprung pogo stick.

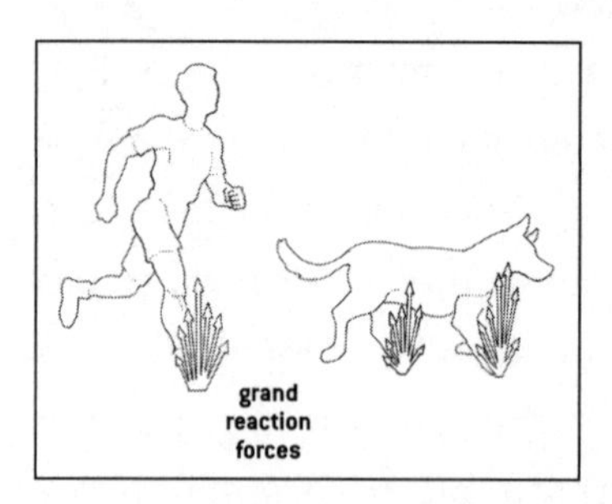

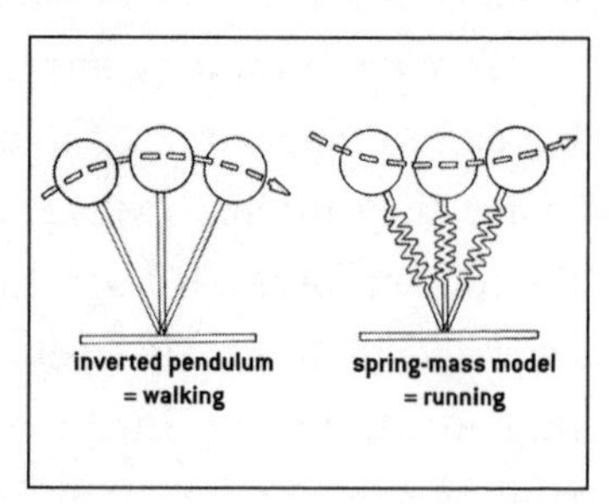

Figure 7.2. Walking versus running. In walking, the leg acts like a rigid pole over which the animal vaults its weight; in running, the leg acts more like a pogo stick.

Fish locomotion is achieved by a lateral body motion imparting energy to the surrounding water to create a chain of interlinked circular vortices, as shown in figure 7.3. Forward motion is then a direct consequence of Newton's second law of motion. In this case, the mathematical equations that govern the motion have hitherto resisted all attempts at mathematical solution. Indeed, it is not even known if there is a solution in the usual sense of a mathematical formula that describes the motion. The illustration of the vortex rings shown in figure 7.3 was generated numerically by a computer. Of course, in the sense of natural computation, the fish "solves" those equations every time it swims.

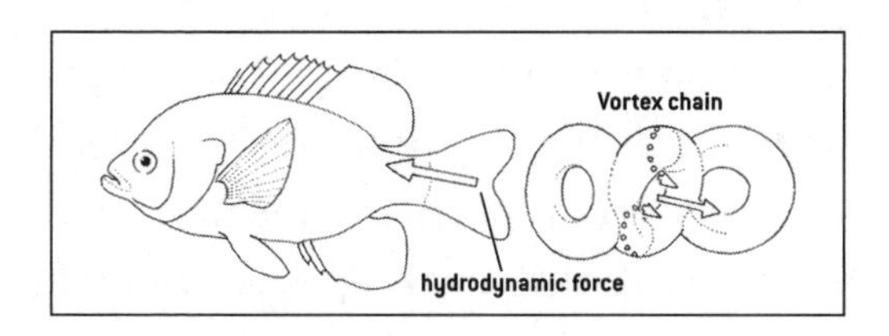

Figure 7.3. A fish moves forward by flapping its tail from side to side. This generates a chain of interlinked circular vortices, drawn here from a computer model. Forward motion is a direct consequence of Newton's second law of motion.

Finally, what about flight? For thousands of years humans have looked up at the birds flying overhead and wondered what it would be like to join them in the air. As we know, it took many years to make that dream a reality, and were able to fly only when we stopped trying to fly the way birds do, by flapping wings, and relying instead on mathematics.

The trick to flying is to be able to exploit Newton's second law of motion in the air. That means creating a downward force (i.e., a downward flow of air) sufficiently strong that the equal and opposite reaction force (known technically as lift) can counter the downward pull of gravity. Helicopters do this directly: the large horizontal rotor blade generates a downward flow of air. In the case of a winged airplane, the downward flow of air is achieved in a more indirect fashion. One or more engines provide the aircraft with a horizontal motion. The motion of the airplane through the air means that, relative to the airplane, the air is flowing backward over the fuselage and the wings. By designing the fuselage and wings to be of a suitable shape, and by angling the fuselage and wings slightly upward (the "attack angle"), the air flowing over the fuselage and wings is forced downward, and the resulting reactive force provides the lift that keeps the aircraft in the air.

As in the case of fish swimming through water, much of the energy imparted to the surrounding air from an airplane flying through it is in the form of vortices in the air. For a large, modern jet airliner, those vortices can trail behind the airplane for several miles and can be sufficiently strong to affect any other airplane that flies through them. As a result, aviation regulations prohibit planes from flying too close together on the same flight path.

The flight of gliding birds can be explained similarly. But the mathematics of birds' flying by flapping their wings is much more complicated. In essence, the motion of

the bird's wings must create a sufficient downflow of air so that Newton's second law provides lift. But exactly how this occurs is not yet known. What we know about how birds fly is more a result of observation of slow motion film of birds in flight, computer simulations, and the construction of physical models than of solving equations of flight.

In the case of many flying insects, and for birds that can hover in one place, such as the hummingbird, the aerodynamics is somewhat different. In such cases, the attack angle of the wing is so steep that the standard mathematics used to describe aircraft flight predicts that the airflow over the wing would break free of the wing surface, causing the bird to stall. What prevents this from happening is that the motion of the wing follows a looped path that produces aerodynamic forces in different directions, but with a net lift. (See figure 7.4.)

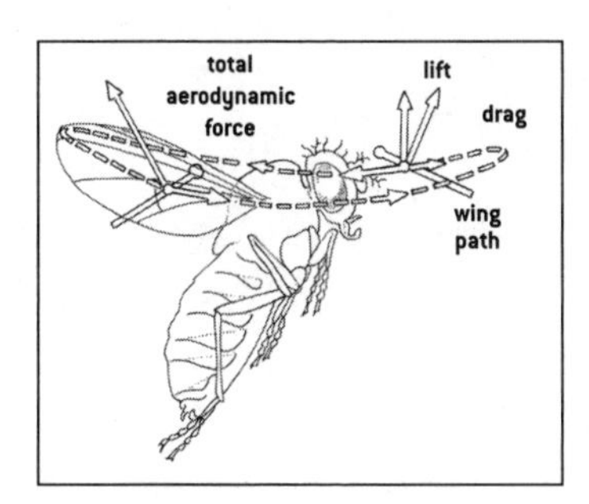

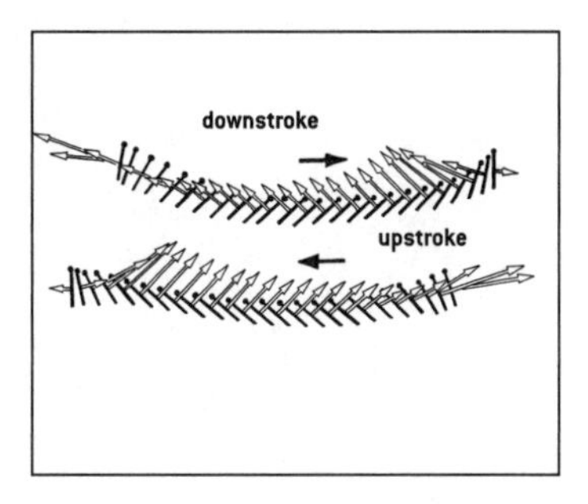

Figure 7.4. For some flying insects and for birds that can hover in one place, the motion of the wing follows a looped path that produces aerodynamic forces generating a net lift.

For smaller flying insects, the mathematics of flight is different again. In such cases, the viscosity of the surrounding air becomes a significant factor, and the insect remains aloft not by forcing air to move downward but by pushing downward with its wings against a body of air that is resistant to being displaced.

Whatever the exact mechanism used to generate lift, however, the important point is that flight involves a lot of mathematics. As usual, nature takes care of this by building the requisite math into the flying creature's structure. But when humans try to do the math explicitly, they find themselves facing equations they are unable to solve, except in an approximate, numerical fashion.

8.

The Eyes Have It: The Hidden Math of Vision

Seeing is so fundamental, something we take so much for granted, something that many living creatures can do, that you could be forgiven for assuming that it is a relatively straightforward process.[23] A simple explanation might say that light enters the eye, is focused by the eye's lens, strikes the retina at the back of the eye, generating an electric current (signal) that travels

23. A good, readable survey of what is known about vision is given in chapter 4 of Steven Pinker's book *How the Mind Works* (W. W. Norton, 1997). Much of this chapter is derived from Pinker's excellent account, to which the reader is referred for further details.

along the optic nerve into the brain, which interprets that signal as seeing. All of this is true, as far as it goes, but that explanation doesn't go very far. In fact, by concentrating on the role played by the eyes, it misses most of the action involved in seeing. For it is not so much our eyes that do the seeing but our brains. (Technically, the eye is a part of the brain, but it is convenient for our present purposes to view it as a separate organ outside the brain but connected to it.) Moreover, the part played by the brain in seeing involves massive amounts of innate mathematics.

A major problem that nature (i.e., natural selection) had to overcome when she designed eyes was making sure that we can see depth, that is, ensuring that we see the world as three-dimensional, full of solid objects, some closer than others, and some partially obscured by others. What makes this a task for the *brain* to perform is that the image created on the retina is two-dimensional–it has to be, since it *is* an image (on a slightly curved, two-dimensional surface). True, in general there are two images, one created in each eye, and that is a large part of nature's way of enabling us to see the world around us as three-dimensional. But binocular (i.e., two-eyed) vision is not the complete solution, since people with only one eye have depth perception ability that is adequate for leading a fairly normal life.

The need for mathematics in vision becomes apparent when you imagine light falling on the retina in the shape of an ellipse (a symmetrical oval). Does this image come from an ellipse viewed head-on, or a circle viewed at a slant? (See figure 8.1. [a]) Suppose you look at a book lying on the table.

Unless you are looking directly down onto the book from above, the shape cast on the retina is actually trapezoidal, if you are looking head-on at one edge, say the bottom of the book (when the nearer edge will create an image on the retina longer than the image from the farther edge, with the images of the two sides both slanted at an angle), or else nonrectangular if the book is at an angle to your position. (See figures 8.1 [b], [c].) Nevertheless, you see the book as rectangular. The brain, when it receives the signal from the retina(s), somehow compensates for the distortions caused to the retinal image(s) by the geometry of the world, in this case the angle at which a circle is inclined to your face or the angle at which the book is slanted toward you.

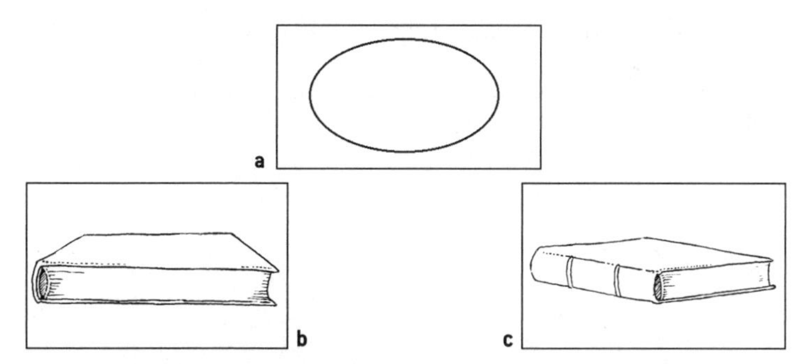

Figure 8.1. (a) An ellipse viewed head-on and a circle viewed at an angle cast identical images on the retina. In the absence of additional cues, if we see an ellipse, we cannot know whether we are looking at an actual ellipse or an actual circle. (b) A rectangular book lying on a desk has a trapezoidal shape when viewed straight on. (c) The same book, when viewed from one side, has a nonrectangular shape. In both cases (b) and (c), the human observer sees the book as being perfectly rectangular; the visual system automatically and subconsciously compensates for the distortion caused by the angle of viewing.

You can think of the problem in terms of what happens to a single photon of light that leaves the object being viewed, passes through the lens of the eye, and then strikes the retina. From a physical standpoint, that photon of light carries no information as to where it started from (more precisely, where it last left a physical surface). It could have emanated a few inches from the eye or many miles away. One way to determine how far away the starting point was (i.e., where exactly was that particular part of the object being viewed?) is to measure the angle between the two eyes when both are focused on that same starting point. This is in fact an important (though not the only) mechanism the brain uses to determine distance, and it requires mathematics—in this case, trigonometry. (See figure 8.2.)

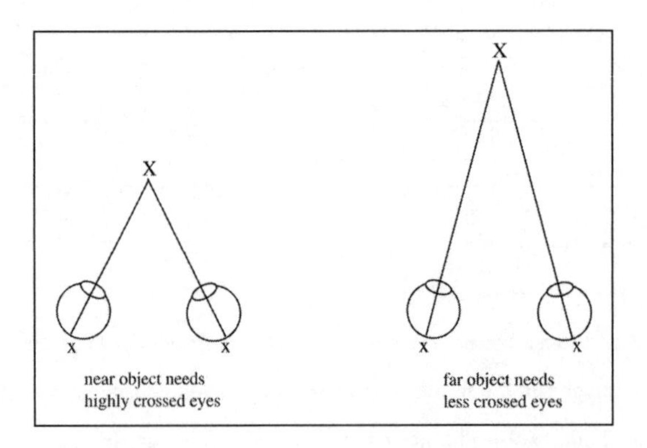

Figure 8.2. The closer we are to an object, the more our eyeballs have to angle inward to focus on it. An elementary trigonometry calculation can determine the distance from the object, given the angle of rotation of the eyes toward the center.

Another way to determine distance is to measure how much the lens of the eye must distort in order to focus on the object. Any lens works by bending the light passing through it so that photons emanating from the same point on the object all converge on the same location on the retina. The closer the object is, the more the lens must bend the light to bring it into focus. The degree to which a lens can bend light is determined (via a fairly sophisticated mathematical formula) by the curvature of its two faces, that is, the degree to which the surfaces curve around into a bowl shape. The lens in the eye comprises a fluid-filled sac whose shape can be altered by muscles. By adjusting the curvature of the surfaces of the lens, the eye can focus on objects at different distances away. The eye muscle-visual system has evolved so that the degree to which the muscle distorts the curvature provides information about the object's distance. (See figure 8.3.)

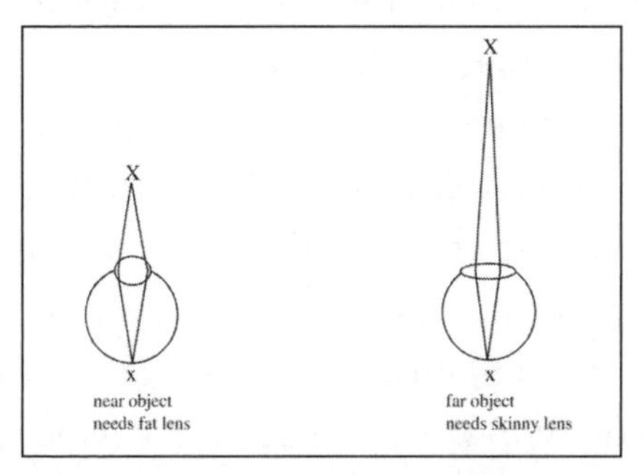

Figure 8.3. The eye focuses on an object by altering the shape of its lens. The closer the object, the greater must be the curvature of the lens's surfaces. A complicated mathematical formula links the distance from the object to the curvature of the lens.

But (innate) mathematics is not enough. Seeing also requires that the mind make various assumptions about what it is seeing, assumptions built up from prior experience of the world. Some of those assumptions come from the environment in which our evolutionary ancestors lived, having been incorporated through natural selection as automatic features of our visual system; others come from our own experience of the world we live in. What we see is conditioned by what was to our ancestors' advantage to see.

This is one occasion where mathematics is not enough. In fact, mathematics plus evolution plus experience isn't enough. The reason is that the problem of determining the true shape of an object from the images it casts on the retinas of our two eyes is simply not solvable. It's unsolvable no matter how many different mechanisms we use. For any image cast on the retinas, there are infinitely many different objects that we *could* be looking at. This is why psychologists and designers of optical illusions can fool us into seeing something that is not really there.

One familiar example is the use of perspective and shading in paintings to give the impression of depth. When we look at a painting or a photograph, we see it as having depth. What we see is not completely three-dimensional. Our minds are not completely fooled. In particular, the eye's lens is able to focus sharply on the entire canvas, telling us that we are looking at a flat surface a few feet away. But at least part of the visual system is

tricked into seeing depth that is not there, and so we get a partial sensation of three dimensions.

A more realistic sense of three dimensions is provided by a device known as the stereogram, which presents the two eyes with separate photographs, taken with two cameras placed side by side the same distance apart as the eyes. (See figure 8.4.) In this case, the illusion of depth depends upon an optical phenomenon known as binocular parallax, first discovered by the physicist Charles Wheatstone in the nineteenth century.

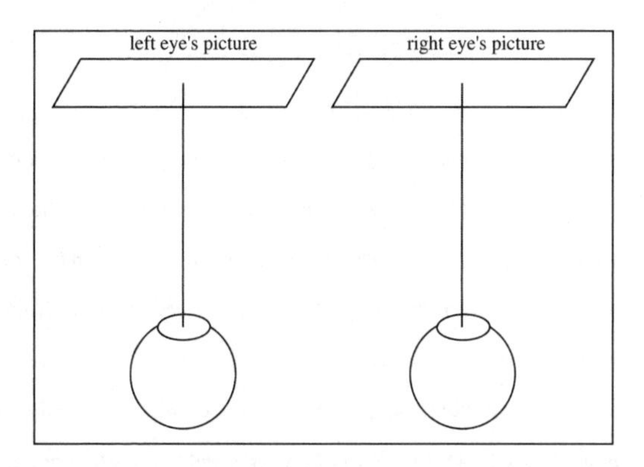

Figure 8.4. The stereogram creates the perception of three dimensions by presenting each eye with a picture of the exact scene it would have received in observing the scene in real life.

The visual system takes advantage of binocular parallax to produce what is known as stereo vision. Stereo vision provides information about the relative placement of objects (i.e., which objects are farther away than which

others, and by relatively how much). Both binocular parallax and stereo vision are illustrated in figure 8.5.

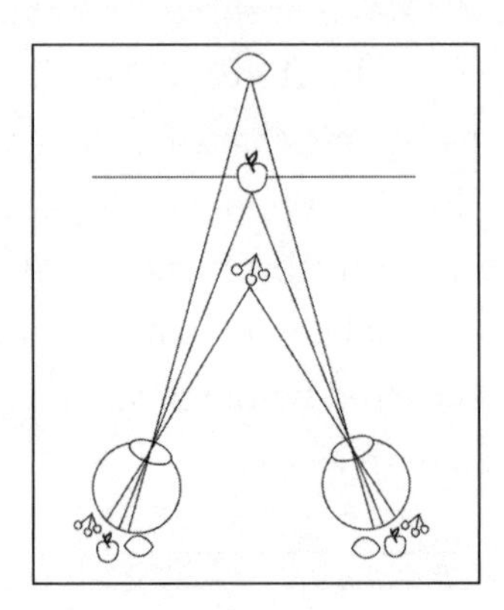

Figure 8.5. Stereo vision. Imagine looking at a table on which some cherries lie in front of an apple, behind which is a lemon. You focus on the apple. The retinal images of the cherries and the lemon are grouped around the images of the apple as shown. The farther away from the apple are the other fruits on the table, the greater will be the separation of the retinal images. The visual system uses this arrangement of retinal images to infer the relative placements of the objects being viewed, and thereby creates a mental image having correct depth placement.

Wheatstone built the first stereogram using wood and mirrors. These days, cheap, commercially produced, plastic stereograms are often sold at vacation resorts, under brand names such as ViewMaster. The ViewMaster is a small boxlike object with two lenses through which the viewer looks at two photographs, often of a local tourist

attraction. Since the photographs are taken with two cameras, placed side by side, the same distance apart as the eyes, the observer looking into the device sees two nearly-identical views of the scene that her two eyes would have seen had she been there in person. A wall that divides the interior of the device into two ensures that each eye sees only the image from the photograph it is supposed to. The reason for the lenses is to compensate for the fact that the photographs are flat images just a few inches from the eye. The visual system has to be fooled (in this case by the lenses) into thinking that the scene being viewed really is some distance away. Wheatstone's original laboratory-built stereogram located the two photographs, which were much larger than those in a ViewMaster, some distance away, using angled mirrors to reflect the images toward the eyes, with a wooden wall separating the two eyes. The distance was sufficient to obviate the need for lenses.

The three-dimensional effect created in the 3D cinemas at amusement parks such as Walt Disney World and at the Kennedy Space Center in Florida is sometimes achieved by the user wearing polarizing goggles to view two images projected using light of different polarizations. Another system has the user wear electronically shuttered LCD (liquid crystal display) goggles that are synchronized to toggle alternately between two images displayed on a computer screen with sufficient rapidity that the visual system cannot detect the changes. In both cases, the idea is to present the two eyes with the images

they would have seen had the observer been present to witness the actual scene. (Of course, what makes this technique particularly effective at amusement parks is that the images can be computer-generated fantasy scenes, thereby providing the observer with a seemingly real visual experience that she or he could never have in real life.)

An early attempt at producing 3D movies was made in the 1950s, when the two stereo images were projected in red and green, and the members of the audience were given cardboard glasses with one red film and one green. The system did produce a stereographic effect, but the overall quality of the film was so poor that the method never really caught on.

Incidentally, in humans, stereo vision and the mechanisms for determining distance are not fully developed at birth, but appear fairly rapidly around three to four months of age. Some of the evidence for this is that, prior to that age, children show little interest in stereograms, but once they are able to perceive the effect, they usually find them fascinating. The standard explanation for this is not that stereo vision has to be learned, but rather that, because it depends on the distance between the two eyes, nature waits until that separation has stopped growing, which occurs around twelve to sixteen weeks after birth. Children or animals who are made to wear a patch over one eye during the crucial stage of development are never able to acquire proper stereo vision or the ability to judge distance.

A more dramatic illustration of fooling the visual systems is provided by autostereograms, those intriguing computer-generated images that initially look like a random jumble of dots or squiggles, but when you stare at them the right way a seemingly genuine three-dimensional picture jumps out of the page or the canvas. They were discovered by accident by the psychologist Christopher Tyler, in the course of his research on binocular vision.

Seeing an autostereogram requires disassociating features of the visual system that normally work together to reduce the likelihood that our minds are being fooled by what we are seeing, which is why it takes most people a few seconds' or even minutes' effort before the effect occurs and the three-dimensional image is formed. Some people claim to be unable ever to succeed, and look on in bewilderment at the "Oohs" and "Ahhs" of those around them who marvel at the experience of seeing "genuine" three dimensions in what they know is a two-dimensional image.

The very first time I saw one of these images was in the early 1990s, when they first hit the markets. I came upon a group of people in a poster shop in Maine, all clustered round this one poster, enthusing at the wonder of what they claimed they were seeing. "Go on, take a look," one of them said to me. I looked. Nothing. "You have to concentrate," said one. "Let your eyes go out of focus," said another. "Look behind the picture," said a third. Still I saw nothing but a random pattern of three colors of dots. After a while, I became convinced I had stumbled into a college

psychology experiment. (This was taking place in a college town and the people in the shop all looked as though they might be students.) I was determined not to say I could see something I could not. I suspected that the experimenters were hoping to show that people are reluctant to admit an inability to do something everyone else can do. But then, after many attempts, I too learned how to let my focus drift back behind the image in just the right way for the three-dimensional image to form–a bit at a time at first, but eventually the whole thing: the Statue of Liberty in that case, one of the first commercially available autostereograms.

Autostereograms work by "tricking" one of the mechanisms that the brain uses to try to determine where the path of a particular photon begins. Our brain assumes that if an image (or part of an image) on one retina looks just like the image (or part of an image) on the other, then both eyes are in fact focused on exactly the same image (or part of an image). Under normal circumstances, this method (illustrated in figure 8.2) works extremely well. The autostereogram works by putting many identical images all over the page, and positioning some (but not all) of them so that the brain thinks certain pairs of images on the two retinas come from the same constituent image when in fact they come from two separate (but otherwise identical) constituent images. (See figure 8.6.)

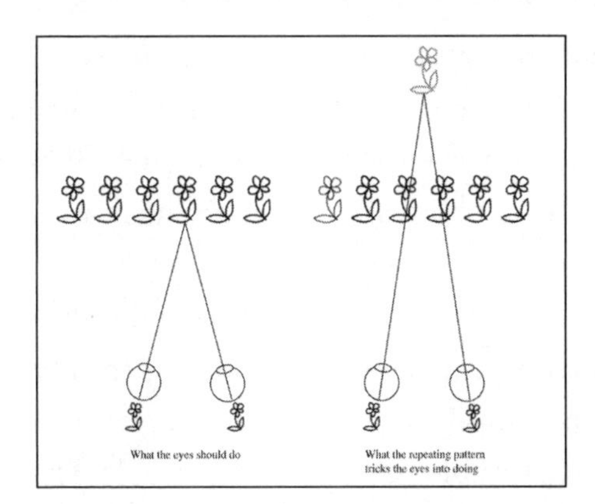

Figure 8.6. (a) The autostereogram fools the visual system into creating the perception of three dimensions. The visual system assumes that identical images in the two eyes come from the same object. A suitably designed pattern of repeating identical picture elements can trick the visual system into regarding two separate elements as a single element behind the canvas, thereby creating a perception of a third dimension.

(b) Stare at this picture face on and the three rows will fall into three distinct layers, the aircraft at the back, the small clouds standing out from the page, and the large clouds farther out still. This is not perspective, but a genuine visual perception of three dimensions.

Yet another way to fool the visual system is to present it with an environment for which neither evolutionary history nor prior experience has prepared it. This is the basis behind those amazing distorting rooms in which a small child can seem to tower over her mother. (You can find them at museums, such as the Exploratorium in San Francisco, or at commercial recreation sites such as the Mystery Spot in Santa Cruz, California.) These are the brainchild of the painter and psychologist Adelbert Ames, Jr. The idea is to build an irregularly shaped room so that, when an observer looks in through a peephole (on whose position the entire design of the room depends crucially), it looks like an ordinary rectangular room. This is done by angling the walls, floor, and ceiling appropriately, drawing lines on the walls, floor, and ceiling (to appear as parallel and perpendicular to one another), and sometimes placing carefully located objects, themselves designed to appear regular though in fact being quite irregular. (See figure 8.7.) When an observer peers in through the peephole, all the visual clues tell him that what he is seeing is a perfectly ordinary room. Thus, the eye-brain visual system processes the scene as if that is exactly what is there. The result is that the observer automatically and subconsciously adjusts the perceived heights of the mother and child to match the environment. Since our minds know that the farther away an object is, the smaller it appears, the daughter, who appears to be much farther away than the mother, but in reality is much closer, is seen as being bigger than her mother.

Figure 8.7. An Ames room. By carefully constructing a highly irregular room so that it looks like an ordinary, rectangular room when viewed through a peephole in one wall, Adelbert Ames Jr. created an environment in which a young child appears to tower over her much larger mother.

We can experience a similar illusion when we look at the moon. When the moon is very low in the night sky, close to the horizon, it appears much bigger than when we see it high in the sky. Obviously, the moon does not change its size depending on where it is in the sky. Rather, when the moon is close to the horizon, the ground provides a distance comparison. Our visual system can tell that the moon is much farther away than anything on the ground, so seeing the moon and the ground together, it automatically adjusts the perceived size of the moon, making it bigger. When the moon is high in the sky, however, our visual system can make no such comparison, and so does not make any adjustment.

Another phenomenon that the visual system uses to help produce three-dimensional vision depends on

motion. Anyone who has seen the movie *Star Wars* (or many like it) or the television series *Star Trek* will know that the simple device of having points of light move out from the center of the screen toward the edges creates a powerful sensation of the viewer hurtling forward through the center of the screen. (Some computer screen savers generate the same effect.) This is because the mind has learned over evolutionary time to interpret outward motion of that kind as motion of the moving objects toward and past the observer. Had our ancestors been constantly presented with such screen-bound moving images as part of their everyday environment, our minds would not have learned to interpret them as necessarily indicating forward movement.

Incidentally, the mathematics that relates outward motion on a flat screen to a sense of three-dimensional flight into the screen (or, in the reverse direction, inward motion of points on a screen to three-dimensional flight out of the screen) is fairly challenging three-dimensional trigonometry (also known as solid geometry). No one would suggest that the mind of a typical *Star Wars* viewer carries out those mathematical computations in any explicit sense. What is going on, once again, is that evolution by natural selection has produced a brain that "does the math" automatically.

Nature has yet more ways to help us see depth. In particular, evolutionary experience has led us to infer (automatically) depth from light source effects and shading,

density and clarity of objects in a scene, perspective effects, and angled edges. Figure 8.8 illustrates these.

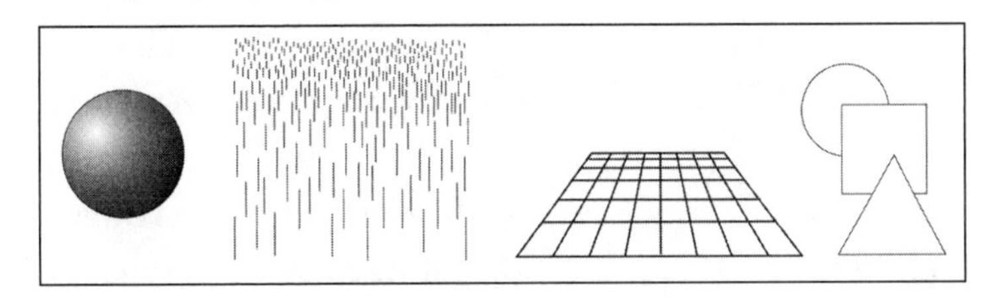

Figure 8.8. Clues our minds use to determine depth in a visual scene (l to r):

(a) Light source and reflection/shadow;

(b) Increasing density and lower resolution;

(c) Perspective geometry;

(d) Occlusion: if one image appears to occlude another, the visual system assumes that one object is in front of the other in the visual field;

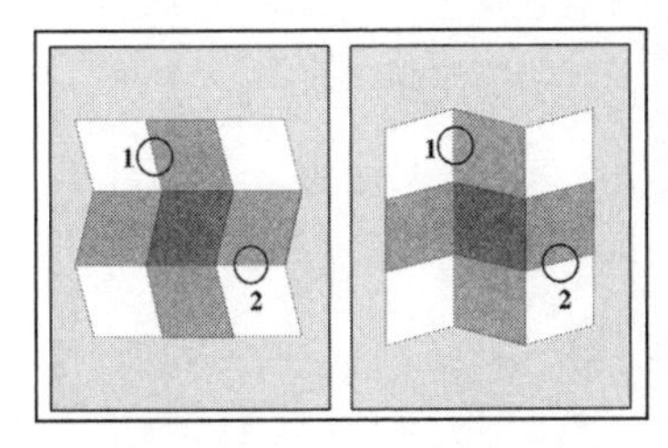

(e) Angled edges: notice how the angling of the edges affects the way we interpret the shading in the two figures, even though, on a square by square basis, the shadings are identical. In the left hand figure, we see boundary 1 as a coloration border and boundary 2 as an angle between two planes; in the right hand figure, we see boundary 1 as an angle and boundary 2 as a coloration border.

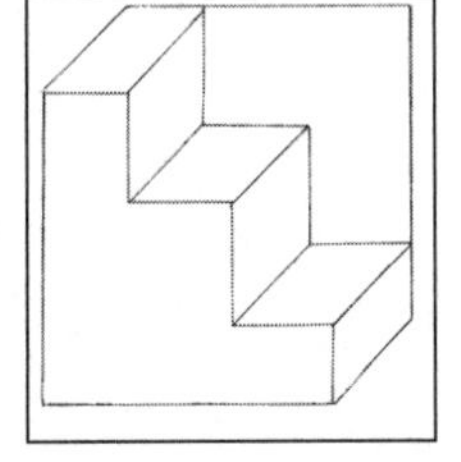

(f) Angled edges: sometimes a two-dimensional image (on its own) is insufficient to determine the three-dimensional object it depicts. We see this image as alternately flipping between a staircase that we are looking down on from above or up at from below.

Finally, there is the puzzling question of how it is that we recognize objects no matter what their orientation toward us. (Authors of illustrated books sometimes manage to find very unfamiliar views of common objects that we have trouble recognizing, but those are generally highly unusual views that would be unlikely to arise in everyday life, and often involve a close-up view of an object generally seen from farther away.) It should be obvious by now why this is a puzzle: the images cast on the retinas are two-dimensional projections of the object. True enough, the mind can then re-create depth using some or all of the tricks we have met. But the result could be a mental image that is quite unlike the more familiar ones of the object concerned. (See, for example, the three views of a suitcase shown in figure 8.9.)

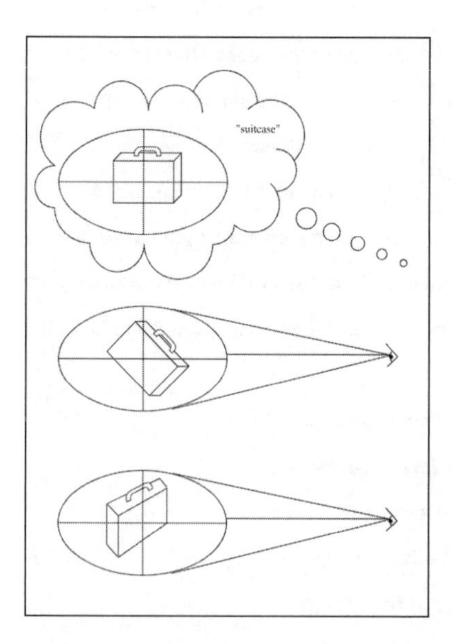

Figure 8.9. Object identification. How do we recognize a suitcase no matter how it is oriented toward us?

Recent research appears to have arrived at the answer. It seems that many objects come with their own preferred frame of reference—their own x, y, z coordinate system if you like. When the mind receives an image of an object, it attaches those coordinates and then mentally identifies the object relative to that preferred frame of reference. (Figure 8.10 illustrates this in the case of the suitcase images from figure 8.9.)

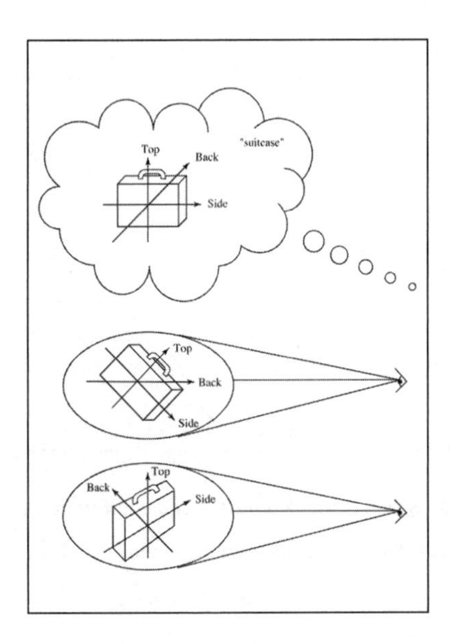

Figure 8.10. Object identification. One current theory suggests that any object has a preferred frame of reference, or axis system, and that we recognize an object relative to its preferred axes. For instance, this figure shows how the suitcase looks the same with respect to its preferred axes, even though it looks different relative to the observer.

This theory suggests that when the mind encounters an object it does not immediately recognize from experience, it mentally rotates it about its preferred axes until it obtains something familiar. Experimental work in psychology indicates that this may well be what is going on,

since the time it takes to recognize an unfamiliar presentation of an object increases in a linear fashion with the number of rotations required to bring it into a familiar position.

You get some idea of how a frame of reference can affect the way we see an object by comparing a square presented with one side horizontal with an identical square rotated through 45 degrees, as in figure 8.11. Everyone sees the first figure as a square and the second as a diamond.

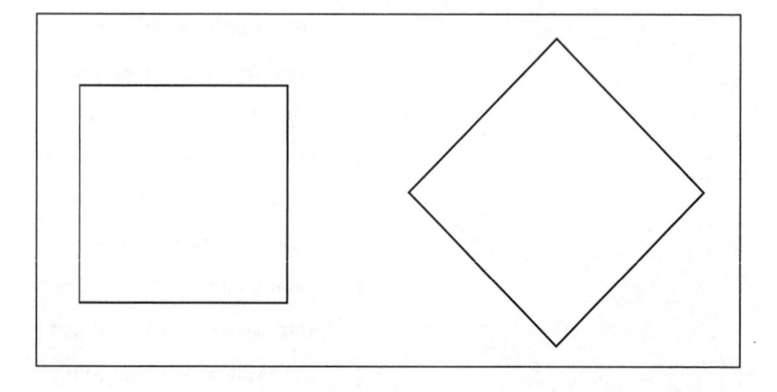

Figure 8.11. Orientation. Identical figures can look different under different orientations. We see the left hand figure as a square, the right hand figure as a diamond, even though the two figures are identical in shape.

Figure 8.12 shows the effect a frame of reference can have. The top right figure looks like a diamond when grouped with the figures to its left, and like a square when grouped with the figures beneath it.

Figure 8.12. Frames of reference. The same object can look very different with respect to different frames of reference. The diagram element in the top right looks like a diamond when grouped with the figures to the left and looks like a square when grouped with the figures beneath it.

It would be possible to list further aspects of vision, but we have now covered the main mechanisms involved in seeing. Certainly, we have seen enough to make it clear beyond any doubt that not only do we see with our minds, but that those minds have to do considerable work to re-create a three-dimensional mental image from the two two-dimensional retinal images produced in the eyes. Moreover, most of the individual techniques used (completely automatically and subconsciously) involve innate mathematics, some of it quite sophisticated. In the sense of abilities that in conscious human terms can only be described as mathematics, it seems that vision is about as mathematical a process as you are ever likely to encounter.

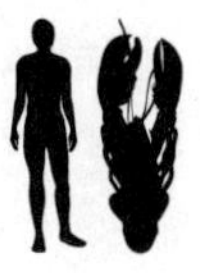

9.

Animals in the Math Class

The examples of animal mathematical feats we have seen so far, while undoubtedly impressive, do not entirely accord with our everyday concept of "doing math." For the mathematics of growth and form (chapter 6), it could be argued that nature *exploited* mathematics in order for animals to develop certain coat patterns or for plants to grow most efficiently. In the cases of the mathematics involved in motion or in binocular vision (chapters 7 and 8), natural selection simply produced creatures whose physical construction *embodied* the appropriate mathematics. This is true even when the creature concerned is us. When our brains create in our minds a three-dimensional image of the world from the two two-dimensional images on the retinas of our two eyes, we are not making any conscious use of trigonometry. We don't have to go to school to learn how to perform this feat. Rather, our brains are simply constructed in such a way that they do this automatically with the signals they receive from the eyes.

On the other hand, the examples we met in chapters 4 and 5, such as the navigation skills of the Tunisian desert ant, the migrating bird, or fish, or the architectural prowess and navigational abilities of the honeybee, definitely involve mental activity. Moreover, when we ourselves carry out that kind of mental activity, we regard it as (possibly subconscious) thinking. But can we really say the creature (including us) is "doing math"? Arguably the "mathematician" to whom the credit should be given here is not the individual ant, bird, fish, or bee, but nature in the form of natural selection. The mental activity of the ant, bird, fish, or honeybee is purely instinctive. In each case, hundreds of thousands of years of evolution have produced a brain that is purpose-built to perform the one or two crucial computations that ensure the creature's survival.

But just because the brain of the desert ant, the indigo bunting, the honeybee, or the seeing human performs certain computations automatically, as a matter of instinct,

does not make the process any less mathematical, or less of a mental feat. After all, we are rightly impressed when a supercomputer solves a difficult equation, and surely right to call it "mathematics," even though the computer is not conscious or in any way aware of what it is doing. So, if we are prepared to concede that computers–completely inanimate objects–can do math, why should we regard similar achievements by living creatures as less worthy of note or of appropriate classification?

Yet you may still believe there is a distinction between natural math and the kind of math we learn at school. So do I. And the difference is just this: the kind of mental processes we usually view as mathematics involves the *mental manipulation* of *numbers*, and other *concepts*. The arithmetical abilities of human babies that we saw in chapter 1 were very definitely of this kind, even though the children themselves were not consciously aware of "doing math." And there is our real question: are humans unique in this regard? Or do any other animals have numerical abilities? Do any other animals have a concept of 1, 2, and 3? Can they do arithmetic? Can they, like humans, *learn* math?

The answer is a definite yes. And we're not just talking about apes and chimpanzees here, our nearest neighbors on the evolutionary tree. Small-brained creatures such as rats and birds also have numerical abilities, which can be improved with training.

RATS!

Numerical abilities have been particularly well documented for rats. (Why rats? Simply this: There is a long tradition of using rats for laboratory experiments in science; they are easily available, easy to handle, and most universities and research labs have facilities that can accommodate them.)

The first convincing evidence that rats have numerical abilities was obtained during the 1950s and 1960s by the American animal psychologist Francis Mechner. In one experiment, Mechner deprived a rat of food for a short period and then put it into a closed box with two levers, A and B. Lever B was connected to a mechanism that delivered a small amount of food. However, in order to activate lever B, lever A had to be first depressed a fixed number of times (n). Moreover, if the rat depressed lever A fewer than n times and then pressed lever B, not only did it not receive food, it also received a mild electric shock. Thus, in order to get food, the rat had to learn to press lever A n times and then press lever B.

With repeated trials, the rats gradually learned to estimate the number of times they had to press lever A before pressing lever B. Thus, if the apparatus was set up so that four presses of lever A were required to activate lever B, then, over time, the rats learned to press lever A about four times before pressing lever B.

It should be noted that the rats never learned to press lever A *exactly* four times on every occasion. In fact, their

tendency was to overestimate, pressing it four, five, or even six times. Given the fact that they received an unpleasant shock if they pressed lever A fewer than four times, this "play safe" strategy makes sense. In any event, it did seem that the rats were able to estimate four presses. Likewise, rats placed in an apparatus set up so that lever A had to be pressed eight times learned to press it about eight times. In fact, positive results were obtained with the apparatus set up for as many as sixteen presses of lever A.

To eliminate the possibility that the rats were judging time rather than the number of presses, Mechner and a colleague, Laurence Guevrekian, carried out a subsequent series of tests, in which they varied the degree of food deprivation. The more hungry the rats were, the more rapidly they pressed the lever. Nevertheless, despite the much faster rate of depression, rats trained to press lever A four times continued to do so, and similarly with the rats trained to another number. Time was not the factor; the rats were estimating the number.

Notice that I keep using the word "estimate." Mechner did not claim that the rats counted. What the experiment showed is that, through training, rats are able to adjust their behavior to press a lever *about* a certain number of times. It may be that they are in fact counting, albeit badly. But there is no evidence for this. It is also possible (and I think far more probable) that they are simply *judging* or *estimating* the number of presses, and moreover doing so

as well as we ourselves could if we did not count. Rats have, it seems, a general sense of number.

A natural question to ask is what evolutionary advantage for rats led to selection for their number sense? What would a rat gain by being able to estimate numbers? One possibility would be the need to remember navigational information, such as that its hole was the fourth one along after the third tree. (In fact, when you stop to think about it, you realize that a basic number sense is extremely useful in finding *your* way around.) It is also useful in keeping track of other animals in the vicinity, be they friends or potential foes.

Birds again

Birds too have been shown to have similar numerical abilities. One of the first researchers to realize this was the German Otto Koehler during the 1940s and 1950s, although for reasons I will explain below, Koehler's observations were not fully appreciated until after Mechner's work.

Koehler showed that birds have both the ability to compare the sizes of two collections presented simultaneously and the ability to remember numbers of objects presented successively in time, both of which are important prerequisites for arithmetic.

In one case, a raven called Jakob was repeatedly presented with two boxes, one of which contained food. The

lids of the boxes had different numbers of spots arranged randomly. A card placed alongside the two boxes bore the same number of spots (though arranged differently) as the lid of the box with the food. Over many repetitions, the raven eventually learned that in order to obtain the food, it had to open the box whose lid bore the same number of spots as the card. In this way, it was eventually able to distinguish 2, 3, 4, 5, and 6 spots.

In another experiment, Koehler trained jackdaws to open the lids of a row of boxes to obtain food until they had taken a given number of pieces of food, say 4 or 5. Each box contained 0, 1, or 2 pieces, distributed randomly on each repetition, so there was no possibility of the birds basing their actions on some other feature such as the length of the row of boxes they opened. Rather, they had to keep an inner tally of how many pieces of food they had taken; in our terms, they had to count the number of pieces of food they had eaten.

Another illustration of birds' numerical abilities comes from Irene Pepperberg, who trained her African Grey parrot Alex to say the number of objects presented to it on a tray, a task that requires that the bird not only distinguish numerosities (or numerousness) but also associate an appropriate vocal response to each number.

Again, many species of birds exhibit number sense in the count of times they repeat a particular note in their birdsong. We know this involves a genuine sense of number because members of the same species of bird born

and reared in different regions will acquire a local "dialect," with the number of repetitions of a particular note varying from one location to another. Thus, although many aspects of a bird's characteristic song may be genetically determined, the number of repetitions of a particular note seems to be acquired by a young bird imitating older birds, most likely its parents. For example, a young canary raised in one area may repeat a particular note six times whereas a canary raised elsewhere will repeat the same note seven times. Since the number of repetitions is constant for each bird, this means that the bird can "recognize" the particular number of repetitions in its song.

HOW MANY LIONS?

As we noted earlier, one obvious survival advantage (and hence a possible evolutionary selection factor) to having number sense–in particular being able to compare numbers of objects in collections–is that it helps a group of animals to know whether to stay and defend their territory or to retreat to safety. If the defenders would outnumber the attackers, it might make sense to stay and fight, but if there are more attackers, the wisest strategy might be to run for it. This suggestion was put to the test a few years ago by researcher Karen McComb and her colleagues. They played tape recordings of lions roaring to small groups of female lions in the Serengeti National

Park in Tanzania. When the number of different roars exceeded the number of lions in the group, the females retreated; but when there were more females in the group than there were different roars, the females kept their ground and prepared to attack the intruders. They seemed able to compare the number of roars that they *heard* with the number of lionesses in the group that they *observed,* a task that requires the abstraction of number from collections encountered through two different senses, hearing and seeing.

Another possible survival advantage to being able to compare numbers of objects in collections (which we observed earlier) is that it is more efficient to expend energy climbing a tree with a lot of fruit than one bearing less fruit.

Beware of the Counting Horse

Now let me explain why Otto Koehler's research on birds was initially not accepted. The story illustrates just how careful you have to be in doing research on mental abilities, particularly with animals.

German scholars in particular were suspicious of claims of mental feats of animals, following the case of Wilhelm von Osten and his horse, Hans. At the start of the twentieth century, von Osten claimed that after ten years of effort, he had succeeded in teaching Hans to do

arithmetic. Both horse and master soon became celebrities, and the German newspapers carried stories about "Clever Hans."

A typical demonstration would see von Osten and his horse surrounded by an eager audience. "Ask him what is three plus five," someone would call out. Von Osten would write the sum on a chalkboard and show it to the horse, who would then carefully tap his hoof on the ground exactly eight times. Other times, von Osten would show Hans two piles of objects, say four in one pile and five in the other. Hans would tap his hoof nine times.

Even more impressive, Hans could apparently add fractions. If von Osten wrote the two fractions $\frac{1}{2}$ and $\frac{1}{3}$ on the board, Hans would tap his hoof five times, then pause, then tap six times, to give the correct answer $\frac{5}{6}$.

Of course, many people suspected a trick. In 1904, a committee of experts gathered together to investigate the matter, among them the eminent German psychologist Carl Stumpf. After carefully observing a performance, the committee concluded that it was genuine—Hans really could do arithmetic.

One person, however, was not convinced by the committee's findings. Stumpf's student Oskar Pfungst insisted on further testing. Pfungst wrote the questions onto the board himself, and he did so in such a way that von Osten could not see what was written. This enabled Pfungst to do something Stumpf had not. On some occasions, Pfungst wrote down the question that had been

given to him. Other times, he changed it. Whenever Pfungst wrote down the question as given to him, Hans got it right. But when he changed the question, Hans gave the wrong answer—in fact, he answered the question von Osten *thought* had been given to the horse.

The conclusion was inescapable: von Osten had been doing the arithmetic. Through some subtle cue, perhaps a raised eyebrow or a slight shrug, he had been instructing Hans when to stop tapping his hoof. As Pfungst acknowledged, von Osten could well have been oblivious to this. Having worked so hard to train him, von Osten very much wanted his four-legged partner to succeed. Possibly he became very tense as Hans's tapping got to the crucial number, and Hans was able to detect some external manifestation of that tension. Thus, while Pfungst's investigation showed that Hans's performance did not require unusual arithmetical powers, it did show that humans may be able to communicate with horses by means of the subtlest actions.

The case of Clever Hans showed the importance of proper design for any psychological experiment, to eliminate any possibility of subtle communication of clues. Unfortunately, the affair made subsequent claims of arithmetical abilities in animals extremely difficult to be taken seriously. And yet nothing Pfungst did showed that animals could not have number sense. He simply showed that *in Hans's case* it must have been von Osten who had performed the calculations, not the horse.

CHIMPANZEES

Turning back to numerical abilities that animals really do possess now, if rats and birds can do some arithmetic, what about chimpanzees? Given their closeness to humans as a species, we might expect that they would exhibit the most well developed number sense. Do chimpanzees in fact have any arithmetical ability? This was the question Guy Woodruff and David Premack of the University of Pennsylvania set out to investigate in the late 1970s and early 1980s.

By most people's standards, Woodruff and Premack started out by aiming high. In their first experiment, the two investigators showed that chimpanzees can understand fractions. For instance, they showed the chimp a glass half filled with a colored liquid and then got the animal to choose between two further glasses, one half-filled, the other three-quarters full. The subjects had no difficulty mastering this task. But was the chimp basing its choice on the volume of the water in the glass or on the fraction by which it was full? The answer to this question was obtained by making the task more abstract. This time, after showing the chimp a half-full glass of liquid, for example, it would be presented with a half an apple adjacent to three-quarters of an apple. The chimp consistently picked the half apple over the three-quarter apple. The same thing happened when the chimp was shown a half a pie against one quarter of a pie. In fact, when presented

with any choice between one quarter, one half, and three quarters, the chimp was able to spot the correct fraction. It knew, for instance, that one quarter of a glass of milk is the same fraction of a whole glass that one quarter of a pie is to a complete pie.

There are many other experiments that have been performed on chimps to show that they possess an ability in basic arithmetic. For example, in an experiment which has been performed on a number of occasions, a chimp is presented with two alternative choices of a treat. On one tray are placed two piles of chocolates, one with three chocolates, the other with four. The alternative tray has one pile of five chocolates together with one additional chocolate in a pile of its own. The apparatus is set up so the chimp can only choose one tray. Which one does it choose? If it bases its choice on the largest pile it sees, it should pick the tray with the pile of five chocolates. But if it can perform the calculation to determine the total number of chocolates on each tray, it will realize that the first tray has a total of seven chocolates whereas the other tray has only six. Most of the time, without any special training, the chimp will select the tray with seven chocolates, showing that it can, in fact, determine which tray has the greater total number of chocolates. In other words, the chimp can, again in the sense of estimation as opposed to exact calculation, perform the additions sums $3 + 4 = 7$ and $5 + 1 = 6$, and moreover can tell that 6 is less than 7.

In many respects, the numerical approximating

ability exhibited by both rats and chimpanzees is similar to the innate numerical estimating ability in humans. But humans can count precisely and are able to perform exact arithmetic by using symbols to denote numbers. Arithmetic can then be performed in an essentially linguistic fashion, by manipulating symbols according to precise rules. A question that naturally comes to mind is: Can chimpanzees be taught symbolic notation?

The answer is yes, up to a point. One of the first successful experiments was in the 1980s, when Tetsuro Matsuzawa, a Japanese researcher, taught a chimpanzee named Ali how to use correctly the nine Arabic numerals: 1, 2, 3, 4, 5, 6, 7, 8, 9. In tests, Ali was able to use these numerals to give the number of objects in a collection presented to him with up to 95 percent accuracy. Ali could recognize at a glance the number of objects in a collection of three or fewer members, but resorted to counting for larger collections. Ali could also order the numerals according to their magnitude.

A number of subsequent investigations have produced similar results. One of the most impressive to date has been the work of Sarah Boyson with her chimp Sheba. Boyson provides Sheba with a collection of cards, on each of which is printed a single digit between 1 and 9. Sheba can correctly match each of the printed digits with a collection of between one and nine objects presented to her. Moreover, the chimp is also able to perform simple addition sums presented to her symbolically. For instance,

Boyson holds up cards bearing the numerals 2 and 3, Sheba can successfully pick out the card bearing the numeral 5.

But, impressive though they may be, are chimpanzees our equals when it comes to numerical ability? Not really. It required many years of slow and painstaking training to achieve the kind of performance exhibited by Sheba and other chimpanzees, monkeys, and dolphins in such experiments. Developing the link between the abstract symbols and the collections of objects they can be applied to is a long and arduous process. Even then, the results are never 100 percent accurate, and they are limited to very small collections. This is quite different from what happens with humans. Young children take just a few months to catch on to numbers. And once they do, they do so in a big–and accurate–way. When it comes to numbers, humans really are very much better than all other animals. And I don't just mean "math types" and "techies," I mean everyone.

Where humans–at least many of us–encounter problems is not with counting but with arithmetic. Now, if we handled larger numbers by simply extending the ability we have for handling small numbers when we are a few days old, it is unlikely, surely, that so many people would come to believe they lack a natural aptitude for math. Presumably, then, we use different methods to perform arithmetic with numbers bigger than 3. What are those methods? We learn some methods in school–at least we are taught them–and we'll come to school mathematics presently. But school is not the only place where people

learn math, and according to the evidence I will present next, it is not the most effective place to learn it. We'll start out with a trip to South America.

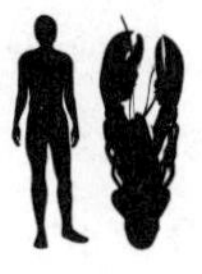

10.

Razor Sharp: The Mathematical Tricks of Street Traders and Supermarket Shoppers

Imagine you are in South America. You are walking through a crowded, bustling, noisy street market. You're actually in the city of Recife in Brazil, but it could be any one of dozens of cities in South America. You walk up to one of the stalls, where a slightly educated twelve-year-old boy from a poor background is selling coconuts.

"How much is one coconut?" you ask.

"Thirty-five," he replies with a smile.

You say, "I'd like ten. How much is that?"

The boy pauses for a moment before replying. Thinking out loud, he says: "Three will be 105; with three more, that will be 210. (Pause) I need four more. That is . . . (pause) 315 . . . I think it is 350."

This exchange is taken verbatim from a report written some years ago by three researchers, Terezinha Nunes of the University of London, England, and Analucia Dias Schliemann and David William Carraher of the Federal University of Pernambuco in Recife, Brazil. The three researchers went out into the street markets of Recife with a tape recorder, posing as ordinary market shoppers. At each stall, they presented the young merchant with a transaction designed to test a particular arithmetical skill.

The purpose of the research was to determine the effectiveness of traditional mathematics instruction, which all the young market traders had received in school since the age of six. How well did our young coconut seller do?

If you think about it for a moment, it is clear that the boy is not doing it the quickest way, which is to use the rule that to multiply by 10 you simply add a zero, so 35 becomes 350. The reason he doesn't do it that way is that he doesn't know the rule. He has never learned it. Despite spending six years in school, he has almost no mathematical knowledge in the traditional sense. What arithmetical skills he has are self-taught at his market stall. Here is how he solves the problem.

Since he often finds himself selling coconuts in groups of two or three, he needs to be able to compute the cost of

two or three coconuts; that is, he needs to know the values $2 \times 35 = 70$ and $3 \times 35 = 105$. Faced with your highly unusual request for ten coconuts, the boy proceeds like this: First, he splits the 10 into groups he can handle, namely $3 + 3 + 3 + 1$. Arithmetically, he is now faced with the determining the sum $105 + 105 + 105 + 35$. He does this in stages. With a little effort, he first calculates $105 + 105 = 210$. Then he computes $210 + 105 = 315$. Finally, he works out $315 + 35 = 350$. Altogether quite an impressive performance for a twelve-year old with little formal education.

But posing as customers was just the first stage of the study Nunes and her colleagues carried out. About a week after they had "tested" the children at their stalls, they went back to the subjects and asked each of them to take a pencil-and-paper test that comprised exactly the same arithmetic problems that had been presented to them in the context of purchases the week before.

The investigators took pains to give this second test in as nonthreatening a way as possible. It was administered in a one-on-one setting, either at the original location or in the subject's home, and included both straightforward arithmetic questions presented in written form and verbally presented word problems in the form of sales transactions of the same kind the children carried out at their stalls. The subjects were provided with paper and pencil, and were asked to write their answer and whatever working they wished to put down. They were also asked to speak their reasoning aloud as they went along.

Although the children's arithmetic was practically faultless when they were at their market stalls (just over 98 percent correct), they averaged only 74 percent when presented with market-stall word problems requiring the same arithmetic, and a mere 37 percent when virtually the same problems were presented to them in the form of a straightforward (symbolic) arithmetic test.

The performance of our young coconut seller was typical. One of the questions he had been asked at his market stall, when he was selling coconuts costing 35 cruzeiros each, was: "I'm going to take four coconuts. How much is that?" The boy replied: "There will be one hundred five, plus thirty, that's one thirty-five . . . one coconut is thirty-five . . . that is . . . one forty."

Let's take a look at this solution. Just as he had done in the exchange described earlier, the boy began by breaking the problem up into simpler ones; in this case, three coconuts plus one coconut. This enabled him to start out with the fact he knew, namely that three coconuts cost Cr$105. Then, to add on the cost of the fourth coconut, he first rounded the cost of a coconut to Cr$30 and added that amount to give Cr$135. He then (apparently, though he did not verbalize this step precisely) noted that the "correction factor" for the rounding was Cr$5, and added in that correction factor to give the (correct) answer Cr$140.

On the formal arithmetic test, the boy was asked to calculate 35×4. He worked mentally, vocalizing each step as the researcher requested, but the only thing he wrote

down was the answer. Here is what he said; "Four times five is twenty, carry the two; two plus three is five, times four is twenty." He then wrote down "200" as his answer.

Despite the fact that, numerically, it was the same problem he had answered correctly at his market stall, he got it wrong. If you follow what he said, it's clear what he was doing and why he went wrong. In trying to carry out the standard right-to-left school method for multiplication, he added the carry from the units-column multiplication (5×4) *before* performing the tens-column multiplication, rather than afterward, which is the correct way. He did, however, keep track of the positions the various digits should occupy, writing the (correct) 0 from the first multiplication after the (incorrect) 20 from the second, to give his answer as 200.

The same thing happened with another child, a girl of nine. When a researcher approached the child at her coconut stall and said "I'll take three coconuts. How much is that?" the young seller replied: "Forty, eighty, one-twenty." With one coconut costing Cr$40, her technique was to keep adding 40 until she reached the correct number of additions.

On the school-style arithmetic test, the same girl was presented with the multiplication 40×3. Her answer was 70. Her is her explanation of how she arrived at that answer: "Lower the zero; four and three is seven."

Clearly, despite the fact that she has no trouble operating a stall in a street market, the young girl's recollections

of the standard arithmetical procedures she has been taught in school are mired in confusion. The same girl, on being asked for twelve lemons, priced at Cr$5 each, separated them out two at a time, saying as she did so: "Ten, twenty, thirty, forty, fifty, sixty." But when she was presented with the sum 12 × 5 on the test–numerically the very same computation–she first lowers the 2, then the 5, and then the 1, giving the answer 152.

A similar degree of confusion about school arithmetic was exhibited by another child vendor who had no trouble with a subtraction task when it arose at the market stall, but went badly astray when presented with the equivalent addition on the school-style written test. Here is the exchange at the market stall, where the boy was selling coconuts for Cr$40 each:

> CUSTOMER: I'll take two coconuts. [Pays with a Cr$500 bill.] What do I get back?
> CHILD: Eighty, ninety, one hundred. Four twenty.

On the test, the child is presented with the addition 420 + 80. He gives the answer 130, apparently proceeding as follows: Add 8 to 2 to give 10; carry the 1; add 1 (the carry), 4, and 8, to give 13; write down the final 0 in the units column to give 130. Eventually, with some prodding by the researcher, the boy was able to reach the right answer–by ignoring the pencil and paper and using a counting method.

A similar outcome arose in another case, after a subject had failed to solve the division problem 100/4. She first tried to divide 1 by 4, then tried to divide 0 by 4, and then gave up, claiming that it was not possible. Prodded by the researcher, she replied: "See, in my head I can do it . . . Divide by two, that's fifty. Then divide by two, that's twenty-five." In other words, she used the fact that dividing by 4 can be achieved by dividing by 2 twice in succession, together with her ability to halve the numbers 100 and 50.

In case after case, Nunes and her colleagues obtained the same results. The children were consistently accurate when they were at their market stalls, but virtual dunces when presented with the same arithmetic problems presented in a typical school format. The researchers were so impressed and intrigued by the children's market-stall performances that they gave it a special name: *street mathematics.*

Street mathematics is the math that people develop for themselves, when they need it. It is not restricted to barely literate market traders in Brazil, and you can find it in other locations besides the streets. For instance, you can find it in the United States, as schoolteacher James Herndon described in his 1971 book *How to Survive in Your Native Land.* Herndon recounts how, on one occasion, he was teaching a junior high school class of children who had all essentially failed in the school system. At one point, he discovered that one of the pupils had a well-paid,

regular job scoring for a local bowling league, a task that required fast, accurate, and complicated arithmetic (have you ever seen the scoring system in bowling?).

Seeing a golden opportunity to motivate this pupil to do well in class, Herndon created a set of "bowling score problems" and gave them to the boy. The attempt was a complete failure. In the bowling alley in the evening, the boy could keep accurate track of eight different bowling scores at once. But he could not answer the simplest scoring question when it was presented to him in the classroom. In Herndon's words, "The brilliant league scorer couldn't decide whether two strikes and a third frame of eight amounted to eighteen or twenty-eight or whether it was one hundred eight and a half."

Herndon met with similar failure when he tried to reach other pupils in the class by presenting them with problems of the very kind they solved with ease outside the classroom. For example, to a girl who admitted she never had any trouble shopping for clothes, he gave the problem: "If you buy a pair of shoes costing $10.95, how much change do you get from a twenty?" (In 1971 this price would have been realistic.) The girl answered "$400.15" and wanted Herndon to tell her if it was right.

Since both the Recife children and Herndon's pupils demonstrated that they could handle arithmetic in certain familiar contexts, when the numbers meant something to them, it seems clear that meaning, or immediate practical significance, plays a major role in our ability to do arithmetic.

But that was not the only difference between street mathematics and school mathematics. The transcriptions of the verbal market stall exchanges showed that the children were using different methods from those taught at school. And yet the school methods are taught because they are supposed to be easier! Indeed, for anyone who masters both methods, the school methods *are* easier–just compare the method our first subject used to compute 10 × 35 with the schoolroom method for solving the same problem. Nevertheless, the people who use street mathematics seem to ignore the standard methods. Why? Intrigued by this question, Nunes and her colleagues set out to examine the methods used by the child vendors.

Their approach was to determine the difference between the children's abilities in mental (or oral) arithmetic and written arithmetic, *when both were measured under test conditions.* As noted already, the children never performed as well when tested as they did when at work at their stalls. But, asked Nunes and her colleagues, was there a measurable difference between the two ways of doing arithmetic on a test? In what ways did the *methods* of street mathematics differ from those of school arithmetic?

The group of children Nunes and her colleagues tested consisted of sixteen students, both boys and girls. All were in the third grade at school, where they had been taught the standard procedures for addition, subtraction, multiplication, and division. Because many children in Brazil have to repeat the same grade level two or more times, the

ages of the children ranged from 9 to 15. The older children had not only had more years' instruction in school arithmetic, they had also spent more time working in the street market.

The subjects were given three kinds of problem: simulated sales transactions of the kind they were familiar with in the market, word problems, and straightforward computational arithmetic problems. In all but one category, the children performed better at mental arithmetic than they did with pencil and paper. In most cases, the differences were dramatic.

Let's start with addition. For the simulated shop questions, the children averaged 67 percent correct orally and 75 percent correct on the written test. This was the only case where their pencil-and-paper results were better than their oral answers (i.e., the answers they obtained by working in their heads without the aid of pencil and paper). For the addition word problems they averaged 83 percent correct orally and just 62 percent written. For the straightforward computation questions, they got a perfect 100 percent orally, compared with a significantly lower 79 percent written.

For subtraction, the difference between their oral performance and their written performance was striking for all three kinds of problem. In the simulated shop they averaged 57 percent correct orally (far less than when calculating change at their stalls) and a mere 22 percent correct written. For the word problems, the figures were 69

percent oral and 22 percent written. For the computation problems, their performances were 60 percent correct orally, but a miserable 14 percent written.

For multiplication, the corresponding figures were a comfortable 89 percent oral and a disappointing 50 percent written for the simulated shop, 64 percent oral and 50 percent written for the word problems, and a perfect 100 percent oral against a poor 39 percent written for the computation problems.

The children scored the lowest on the division problems. They averaged 50 percent oral on all three kinds of problems, but they had clearly failed utterly to master the schoolroom method of division. When asked to answer the questions using pencil and paper, they scored zero percent on the simulated shop and the word problems, and got just 7 percent correct on the straightforward division questions. In short, the children could not do division under any sort of test conditions.

Clearly, the children were much better at mental arithmetic than they were at applying the paper-and-pencil methods they had been taught at school. And presumably the same will be true of anyone who makes regular practical use of numbers and basic arithmetic. But there is still the question of how they were so much more successful in oral arithmetic than in written arithmetic. Since they appeared unable to use the methods they had been taught in school, just how were these children solving the problems when they worked them in their heads?

You get some idea of the children's methods–and hence a first indication that street mathematics is something very different from school arithmetic–when you look at the transcripts of what the children actually said as they were working out the problems mentally. Their words reveal that they were using some sophisticated manipulations of numbers.

For example, when faced with computing 200 - 35, one child proceeded like this:

> If it were thirty, then the result would be seventy. But it's thirty-five. So it's sixty-five. One hundred sixty-five.

Let's look at his method. First he splits the 200 into 100 + 100. (He doesn't vocalize this step, but it's clear from what comes later that this is what he did.) He puts one 100 to one side and sets out to compute 100 - 35. To do this, he first rounds off 35 to 30, and computes 100 - 30. This he can do easily: the answer is 70. Then he corrects for the rounding by subtracting the 5 he ignored: 70 - 5 = 65. Finally, he adds the 100 he had put to one side at the beginning: 65 + 100 = 165.

Even more impressive for its numerical versatility is the method another child used to compute 243 - 75, presented to him as a shopping transaction involving calculating change. Here is what he said:

> You just give me the two hundred. I'll give you twenty-five back. Plus the forty-three that you have, the hundred and forty-three, that's one hundred and sixty-eight.

Faced with a young child at a stall in a busy South American street market who calculated our change in that fashion, many of us would suspect that our young salesman was trying to pull a fast one on us. But this boy's answer was perfectly correct. Here is what he was doing.

First, it is clear from what he went on to say that his first sentence was meant to be "You just give me the one hundred." What he is doing is splitting up the 243 into 100 + 100 + 43. He puts the 43 and one of the 100s to one side and subtracts the 75 from the remaining 100. That's something he can do easily: 100 - 75 = 25. Then he adds back the 43 and the 100; to do this, he first computes 100 + 43 = 143 and then calculates 25 + 143 = 168. That last step is still a challenging addition, of course. In essence, what his overall method does is change a challenging subtraction problem, 243 - 75, into a less challenging (but still difficult) addition problem 143 + 25. This works because, like most people, he finds addition much easier than subtraction.

Let's look at one more example, this time involving division. As we saw earlier, most of the children had significant difficulty with division when working orally

and failed completely when trying to use the school procedure. The problem was to calculate 75/5, asked as a question about sharing 75 marbles among 5 children. Here is what one child said:

> If you give ten marbles to each, that's fifty. There are twenty-five left over. To distribute to five boys, twenty-five, that's hard. . . . That's five more for each. Fifteen each.

That's correct. The child begins by "rounding" 75 to 50 and solving the simpler problem 50/5, for which he has no trouble computing the answer 10. (Indeed, he presumably knew that, which was why he performed the initial rounding down from 75 to 50.) The rounding leaves 25 marbles still to distribute. He finds this difficult; he does not know the answer to 25/5. But after a bit of thought he figures it out: 25/5 = 5. Now all he has to do is add that 5 to his previous result of 10 to give his final answer, 15.

Faced with the evidence of the Brazilian street traders, most people concede that if they ever found themselves in a position where their very livelihood depended on math skills, they probably could acquire them. But once you admit that, then you've acknowledged that the only thing that stands in the way of your improving your math skills is lack of motivation and practice.

Math whizzes with shopping carts

Despite their failure to master school mathematics, the child vendors in Brazil and the other groups who used street mathematics all had one thing in common: they used numbers frequently in a context where those numbers had immediate practical significance. For most of us, that's not the case. Most of the time we can get by just fine without using arithmetic. But one situation where practically all of us encounter numbers is shopping. To be sure, even for the most careful shopper, the use of arithmetic is far less repetitive and much less intense than it is for a market vendor. Moreover, no matter how concerned we might be with getting the best buy at the supermarket, there is far less pressure to get the right answer than there is for a market trader whose livelihood is at stake. So there is no reason to expect the average price-conscious supermarket shopper to exhibit the numerical dexterity of those Brazilian street vendors. But just how much arithmetic do we use, and how well do we use it?

This was the question addressed some years ago by the anthropologist Jean Lave in a study called the Adult Math Project (AMP). Currently a faculty member in the Department of Education at the University of California at Berkeley, Lave was at the University of California at Irvine at the time of the study. The subjects she studied were ordinary people in southern California, shopping in a supermarket.

Lave's study differed from that of Nunes and her

colleagues in one important way. In the case of the young Brazilian street traders, the unusual ad hoc methods of computation they adopted were generally entirely mathematical, so it was possible to evaluate them from a purely mathematical standpoint, from which perspective they actually appear quite sophisticated. But in many of the cases cited in the Adult Math Project, the shoppers used a combination of mathematical and other kinds of reasoning, which the investigators could not evaluate using purely mathematical criteria.

This point is well illustrated by a case from another of Lave's studies, which looked at the mathematics used by dieters when they prepared their calorie-controlled meals. A male dieter preparing a meal was faced with having to measure out $3/4$ of the $2/3$ of a cup of cottage cheese stipulated in the recipe he was using. Before reading on, ask yourself, how would you go about this?

Here is what this man did. He measured out $2/3$ of a cup of cheese using his measuring device, and spread it out on a chopping board in the shape of a circle. He then divided the circle into four quarters, removed one quarter and returned it to the container, leaving on the board the desired $3/4$ of $2/3$ a cup. Perfectly correct.

What is my reaction as a mathematician? That there is a much easier way: By cancellation (of a common 3) followed by simplification (dividing by a common factor of 2), you get

$$3/4 \times 2/3 = 2/4 = 1/2$$

So all the subject needed was ½ a cup of cheese, which he could have measured out directly. Simple. But our man did not see this solution. Nevertheless he clearly knew what the concept "three-quarters" means, and was able to use that knowledge to solve the problem in his own way. He got the job done, and in Lave's terms he successfully solved the problem.

WHAT SUPERMARKET SHOPPERS GET RIGHT

The AMP studied twenty-five shoppers in Orange County in southern California. Orange County is commonly regarded as a highly affluent region with very conservative politics, but the subjects varied considerably in terms of education and family income, and included some people of poor educational background and low income for whom thrifty shopping was essential.

Since the idea of the study was to examine the way ordinary people actually used mathematics in their everyday lives, the researchers could not simply test them with questions such as "If you were faced with three kinds of frozen french fries with the following weights and prices, how would you decide which is the most economical?" As we'll see below, the answer that people give to such a question has very little to do with what they would actually do in a real shopping situation. "What if?" questions don't work.

Instead, the researchers chose to follow the shoppers around and observe them, taking notes, occasionally asking them to explain their reasoning out loud as they went about their shopping, and sometimes asking for explanations just after the transaction had been completed. Of course, this procedure is highly contrived, and the very presence of an observer changes the experience of shopping. Thus, to some extent the study is not really one of people "in their normal, everyday activities." But it is probably as close as you can get. Moreover, anthropologists have developed ways of going about such work so as to minimize the effect of their presence on their subjects' behavior.

Each researcher spent a total of about forty hours with each of her or his subjects, including time spent interviewing them to determine their backgrounds (education, occupation, etc.). Although most of the shoppers were women, there were some men in the group. The researchers noticed no difference in the mathematical performances of the men and the women in the supermarket, however, so gender did not seem to be a significant factor.

Out of a total of around eight hundred individual purchases that the shoppers made in the course of the study, just over two hundred involved some arithmetic, which the researchers defined to be "an occasion on which a shopper associated two or more numbers with one or more arithmetical operations: addition, subtraction,

multiplication, or division." The shoppers varied greatly in the frequency with which they used mathematics. One shopper used none whatsoever, while three of the subjects performed calculations in making over half their purchases. On average, 16 percent of purchases involved arithmetic.

One interesting finding was that, in comparing competing products to decide which was the best buy, shoppers made relatively little use of the unit price printed on the label–information included on the label by law specifically to enable shoppers to compare prices. The investigators were not entirely certain why this was the case. The most likely explanation they could offer was that the unit price is essentially abstract, arithmetical information. Unless the product is something that the shopper either buys or uses in, say, single-ounce units, then the price per ounce has no concrete significance for that shopper. Thus, even though direct comparison of unit prices is the most direct way to determine value-for-money, shoppers often ignore it.

A common approach was to calculate ratios between prices and quantities in a way that made direct comparison possible. This could be done if the quantities were in a simple ratio to one another, say 2:1 or 3:1. For example, if product A cost $5 for 5 oz. and product B cost $9 for 10 oz., the comparison was easy. A typical shopper would reason like this: "Product A would cost $10 for 10 oz., and product B is $9 for 10 oz., hence product B is the cheaper

buy." Shoppers would often abandon the comparison approach when faced with a ratio of 3:2, where making a comparison would require multiplying one price by 2 and the other by 3.

Another advantage of working with the actual amounts that might be purchased–rather than with the more abstract figures of unit prices–is that the price comparison is often just one part of a more complex decision-making process in which the shopper's storage capacity, size of family, likely rate of usage, and the estimated storage period before a particular item might spoil all play a part. As the researchers observed time and again, what shoppers did was to juggle all of these variables in order to reach a decision, thinking about the purchase options first one way, then another, then another. The price-comparison arithmetic was only one part of this process. Despite the complexity of this process, shoppers did not have to expend any great effort. Indeed, they were not aware that they were "thinking" much at all; they were "just shopping."

A slightly different kind of transformation to facilitate a price comparison involves conversion of units. For instance, Lave quotes the following exchange between an AMP shopper and her daughter:

> Daughter: 18.
> Shopper: 18 ounces for 89, and this is? [Refers to another brand]

Daughter: One pound, seven ounces.
Shopper: 23 ounces for a dollar 17.

After converting the weight of the second brand from pounds and ounces to ounces, the shopper was faced with a weight ratio of 18:23, and at that point she abandoned that approach and based her decision on other factors. Indeed, when faced with a particularly problematic comparison in the supermarket, when the units could not easily be made to match, shoppers would often give up the attempt and base their decision on some other consideration—perhaps choosing the larger quantity because bigger quantities are generally more economical. From the point of view of "doing math," abandoning the calculation leaves the problem unsolved, of course. But that does not mean that the overall mental process was a failure. After all, people don't go into a supermarket in order to carry out arithmetical calculations, they go there to shop, and from the point of view of a shopper trying to make prudent purchases, arithmetic is just one of several strategies that can be used. Thus, even when a subject is unable to do the necessary arithmetic, by whatever means, the shopping expedition could still have been a success in terms of making sensible purchases.

Another method that many shoppers used to decide between two options was to calculate the price differential, a procedure that requires just two subtractions. For example, faced with a choice between a 5-oz. packet

costing \$3.29 and a 6-oz. packet priced at \$3.59, one shopper reasoned, "If I take the larger packet, it will cost me 30 cents for an extra ounce. Is it worth it?"

Among the arithmetical techniques that the researchers observed shoppers performing were estimation, rounding (say, to the nearest dollar or the nearest dollar and a half), and left-to-right calculation (as opposed to the right-to-left calculation taught in school). What seemed to be absent, however, were most of the techniques the shoppers had been taught in school. Lave and her colleagues set out to investigate where the school math had gone.

In order to compare the shoppers' arithmetical performance in the supermarket with their ability to do "school math," the researchers designed a test to determine the latter. Again, the findings were interesting. Despite the significant efforts the researchers made to persuade the subjects that this was not like a school test, but that its purpose was purely to ascertain what arithmetical ability they had retained since school, the shoppers treated it as if it were indeed a school test. For instance, when the researchers asked if they might observe the subjects taking the test, the subjects responded with remarks such as "Okay, teacher." They made comments about not cheating. They asked if they were allowed to rewrite problems, and they spoke self-deprecatingly about having not studied math for a long time. In other words, the subjects approached the math

test in "math test mode," with all the stresses and emotions that usually entails.

Perhaps the shoppers' reaction was to be expected. After all, the "math test" did have all the elements of a typical school arithmetic test: questions involving whole numbers, both positive and negative, fractions, decimals, addition, subtraction, multiplication, and division. On the other hand, the problems were designed to test the same arithmetical skills that the researchers had observed the shoppers using (in context) in the supermarket. For instance, having observed that shoppers frequently compared prices of competing products by comparing price-to-quantity ratios, the researchers included some problems to see how the subjects fared with abstract versions of such problems. For example, faced with an item costing $4 for a 3-oz. packet and a larger packet costing $7 for 6 oz., many shoppers would in effect compare the ratios $4/3$ and $7/6$ to see which was the larger. So the researchers would include on the test the question: "Circle the larger of $4/3$ and $7/6$." But the same shopper who did just fine in the supermarket would fail on the test.

Overall, the shoppers' performance was rated at an average 98 percent in the supermarket compared to a mere 59 percent average on the test. Why? One obvious difference was that the subjects took the test questions as requiring a precise calculation, whereas they were much more likely to use estimation in tackling their real-life equivalents. The main difference, however, was that the

shoppers in the supermarket were not using the arithmetical skills they learned in school. Rather, they were solving the problems another way.

That last conclusion is supported by the fact that performance on the test was higher the longer subjects had studied math at school and the more recently they had finished school, whereas neither length of schooling nor the time since schooling had any measurable effect on how well they did in the supermarket. Thus, if they teach anything, school math classes seem to teach people how to perform on school math tests. They do not teach them how to solve real-life problems that involve math.

We'll come back later to the question of why school math classes don't seem to achieve the goals they are supposed to–and what we might do to improve matters. In the meantime, let's take a look at the problems that caused the supermarket shoppers the greatest difficulty.

WHAT SUPERMARKET SHOPPERS GET WRONG—AND WHY

The shoppers Jean Lave studied in the Adult Math Project were highly successful when it came to performing arithmetical tasks in real-life situations, regardless of their schooling history. How were they doing it?

Of course, part of the difference in performance might have been due to the difference between actually being in

the store as against "taking a test." As we have seen, the subjects could not help viewing the arithmetic test as a "school quiz." But that did not seem to be the major factor. Rather, what seemed to make the biggest difference was the kind of test the shoppers were asked to take and the manner in which the questions were presented. This was shown by a further test the AMP researchers put the subjects through: a shopping simulation.

In their homes, the subjects were presented with simulated best-buy shopping problems, based on the very best-buy problems the researchers had observed them resolve in the supermarket. In some of these simulations, the subjects were presented with actual cans, bottles, jars, and packets of various items taken from the supermarket and asked to decide which to buy among competing brands; in others they were presented with the price and quantity information printed on cards. In this simulation, which was evidently a kind of "test" situation, the subjects scored an average of 93 percent. (The fact that the simulation was done in the subject's home, carried out by the researcher who had accompanied the subject on the shopping trip, also seems to have been a significant factor. I'll come back to this point in a moment.)

To put this in terms of a specific example, a subject would perform extremely well (in the 93 percent success rate category) in the home shopping simulation, when presented with a card that said 3 oz. of product A cost $4 and another card that said 6 oz. of product B cost $7 and asked

which was the best buy; but in the context of being presented with a list of arithmetic problems, the same subject would do far worse (in the 59 percent success rate category) when asked to circle the larger of $4/3$ and $7/6$. And yet, the arithmetic problem underlying the two questions is exactly the same!

The conclusion seems to be not that people can't do math; rather they can't do *school* math. When faced with a real-life task that requires arithmetic, most people do just fine—indeed, 98 percent success is virtually error-free.

In a moment, we'll see just what kinds of arithmetical problem cause ordinary people most trouble, and ask ourselves why. But before we do, let me mention that, although a number of the AMP shoppers carried a calculator with them, only on one occasion during the entire project did one shopper take it out and use it in order to carry out a price comparison. And no one ever used a pencil and paper to carry out a calculation.

Now let me say a bit more about the nature of the shopping simulation "test" that the AMP researchers carried out in the shoppers' homes. As we have seen, the subjects, performance on the simulation was almost as good as when they were actually shopping. Almost certainly this was because, not only did they not view it as a "math test," they actually managed to approach most of the questions using the same mental resources they had used in the store. The researchers went to some effort to achieve this, giving the subjects the questions verbally in a conversational fashion

and making frequent references to the actual shopping expedition the two had gone on together.

The importance of setting up the shopping simulation test this way becomes clear when you compare the results with those from another "shopping simulation test," carried out by Deanna Kuhn.

Kuhn set up a table outside a southern California supermarket, stopped customers about to enter to do their shopping, and asked them to calculate which of two bottles of garlic powder was the better buy, the 1.25 oz. bottle for 41 cents or the 2.37-oz. bottle for 77 cents, and similarly for two bottles of deodorant, one costing $1.36 for 8 oz., the other $2.11 for 12 oz. The subjects were given a pencil and paper on which they could do their work.

The results were very different from those obtained in the AMP shopping simulation. Only 20 percent of the fifty shoppers who agreed to take the test were able to solve the garlic powder question, with its difficult weight ratio of 1.25:2.37, and not a great many more–just 32 percent–could solve the deodorant question, where the weight ratio is 2:3.

The enormous difference between the results of the two test procedures, AMP and Kuhn, is almost certainly due to the way the subjects approached the two simulations. In the AMP simulation, the shoppers seemed to understand that they were to imagine they were actually shopping, whereas Kuhn's subjects seemed to view it as "taking a test." In fact, Kuhn's results were very similar to

those obtained in the school-style tests administered to the AMP subjects.

In other words, you can carry out the test outside a supermarket and phrase the questions in terms of shopping, even to the point of presenting the subjects with actual items taken from the supermarket shelves, but if the subjects view it as a "math test," that is how they will approach it. As a result, they will struggle to use their long-forgotten (and possibly never fully understood) school math procedures. And more often than not, they will fail.

What problems caused test subjects most difficulty on the formal math test? Not surprisingly, division gave rise to a lot of trouble. Many people gave the wrong answer (or failed to complete) the questions $.7\overline{)1.47}$ and $.6\overline{)24}$. On the other hand, the success rate was higher for the divisions $5\overline{)3.55}$, $26\overline{)100}$, $8\overline{)124}$, and even for $24\overline{)984}$, so the difficulty in the former examples seems to be caused by the decimal point in the divisor.

Decimal points cause problems in multiplication as well. Ending up with the decimal point in the wrong place caused many subjects to give the wrong answer to the multiplications $.42 \times .08$ and $3.5 \times .6$.

The placement of decimal points can also give rise to difficulties when it comes to subtraction, unless the point occupies the same position in the two numbers at the start. For instance, people did reasonably well on the subtraction .81 -.05, where the decimal point is in the same

position in both numbers. But they had trouble with 3.75 - .8 and with 6 -.25. (In fact, this last one seemed to cause grief for surprisingly many people, despite the fact that all it requires is for the subject to subtract $\frac{1}{4}$ from 6.)

Most people were able to handle addition of decimal numbers, and they didn't break a sweat when faced with addition or subtraction of whole numbers, but addition and subtraction of fractions were real killers. Additions such as $\frac{1}{5} + \frac{2}{3}$, $\frac{1}{2} + \frac{5}{6}$, and $5\frac{1}{3} + 4\frac{3}{4}$ proved a major challenge, as did subtractions such as $\frac{3}{4} - \frac{2}{3}$, $\frac{3}{5} - \frac{1}{10}$, and $3\frac{1}{3} - \frac{1}{2}$.

Division of fractions–with the complicated rule of "invert the divisor and multiply"–also caused trouble, even for supposedly easy examples such as $8 \div \frac{1}{2}$, let alone "harder" ones like $\frac{3}{2} \div \frac{1}{4}$ or $\frac{2}{3} \div \frac{4}{5}$.

Perhaps the most surprising result from the AMP math test was the number of subjects who gave the wrong answer to the multiplication $16 \times \frac{1}{2}$, even though they did fine with the problems $\frac{2}{3} \times \frac{5}{7}$ and $\frac{4}{5} \times \frac{3}{4}$.

At first glance, it is not obvious what makes some problems harder to solve than others. But there is a pattern. All the questions on which subjects performed reasonably well could be solved exactly as they were presented. The problems that caused difficulty all required either some initial transformation before they could be solved, or else, in the case of decimal multiplication, a final transformation to position the decimal point correctly. For instance, addition or subtraction of fractions requires an initial transformation to make the

denominators the same (for example, writing $^1/_5 + ^2/_3$ as $^3/_{15} + ^{10}/_{15}$), and division of fractions requires an initial inversion of the divisor (for example, transforming $^3/_2 \div ^1/_4$ to $^3/_2 \times ^4/_1$). Even the simple-looking $16 \times ^1/_2$ must first be transformed to $16 \div 2$.

The principal difficulty with school arithmetic, it seems, is not the *basic* additions, subtractions, multiplications, or even divisions, but the transformations that often have to be made before or after those basic arithmetical steps are carried out. This hypothesis was confirmed by an additional test the AMP researchers asked the subjects to take, which showed that their knowledge of basic additions, subtractions, multiplication, and divisions of pairs of one-, two-, and even three-digit positive whole numbers was fine. For example, subjects had no difficulty computing sums such as $12 + 9$, $31 - 11$, 7×12, or $72 \div 9$.

To summarize, the reason people have difficulty with school arithmetic seems to be that they emerge from schools either having failed to master, or having only partially understood, the all-important transformation rules.

This observation is particularly intriguing in view of the fact that shoppers in the supermarket seem to perform almost all of their numerical problems by a series of transformations that change the problem into an equivalent one that they find easier, often avoiding any actual computation at all (in the usual sense of the word "computation"). Since the transformations shoppers perform correctly in determining best-buys often correspond to the

transformations that should be carried out to solve the equivalent school-math problem, the most likely explanation of the disparity is that the students learn the transformation procedures taught at school *by rote,* without ever achieving any real understanding. But as soon as that student becomes an adult shopper, he or she has little trouble developing the ability to apply the same transformations to real-life situations.

The difference between performance on school-style tests and the use of arithmetic in real life appears particularly dramatic when you realize that the AMP subjects were devoting their full attention to the problems on the school-style arithmetic test, and getting them wrong, whereas the (equivalent) arithmetic they performed with near perfect accuracy while shopping was done while they were engaged in another activity involving various other thought processes and while subject to distractions and interruptions.

Similar findings have emerged from other studies by other researchers. To give just one, there was a study of the employees in a dairy who loaded the delivery trucks. In their daily work, the loaders made virtually no mistakes in computing the amount loaded, despite the fact that some products had 16 cartons to a case, some 32, and some 48, with some cases being sent out full, others only partially full. And yet, when presented with a school-like test of exactly the same arithmetical tasks, the dairy loaders scored a paltry 64 percent on average. Although the

amount of schooling the dairy workers had completed did influence their scores on the school-like tests, it did not affect their arithmetic performance on the job. Indeed, some of the loaders had not finished elementary school, but their arithmetical performance at work was as good as anyone else's, even though it often involved arithmetical tasks beyond anything they had ever encountered in the classroom. Moreover, the more years the loaders had worked in the dairy, and consequently the longer since their math lessons in school, the better their on-the-job arithmetical skills became.

Again, the evidence from the way people handle numbers in the course of their everyday personal or professional lives shows that arithmetic instruction in school does not appear to have the effect that most people assume is the intention, namely, to provide mastery of efficient means of doing arithmetic. This is not to say that such instruction is a waste of time, or that it does not lead to better numerical skills. I think that confusion of these issues lies behind many of the heated arguments about mathematics education that preoccupy parents and school boards.

What the evidence surely does tell us, however, is that if we want to improve our chances of learning math, we need to take a long, hard look at the context and manner in which mathematics is presented. We will return to this topic in the final chapters of the book. But first, let's look at how it is that humans are able to outperform all other

animals in numbers and arithmetic. If you are one of the many people who excuse a middling performance at arithmetic with the fact that you are really quite good with language, then you may be in for a surprise, for the key mental capacity that enables humans to do arithmetic is our capacity for language.

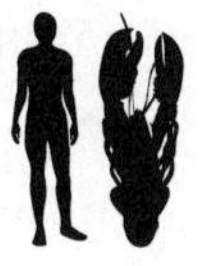

11.
All Numbers Great and Small

As we saw in chapter 1, human beings are born with (or acquire automatically soon after birth) a number sense that allows them to distinguish one, two, and three objects or sounds. By the age of four months, they know (perhaps unconsciously) that, when you take two individual objects, the result is a collection of two objects, not one or three. They know that when you remove one object from a collection of two, you are left with one object, not two objects or none.

Surprising though it may seem, these abilities are not

unique to humans. Using similar techniques to those used with small children, animal psychologists have shown that rats, various kinds of birds, lions, dogs, monkeys, chimpanzees, and other animals have a similar number sense. But there are some significant differences. One difference between humans and animals in terms of numerical ability is that, even at a very early stage, human babies outperform all other species in terms of accuracy.

Another way humans perform much better than all other species is by being able to go beyond our basic one, two, three number sense to handle much larger numbers. But we do that by adopting a very different method, based on counting. That approach uses different mental abilities located in a different part of the brain from the number sense.

How do we count?

The ability to count seems almost, but not quite, uniquely human. With massive amounts of training, scientists have managed to teach chimpanzees, apes, and some kinds of monkey to be able to count up to around 10 or so, with a reasonable (though by no means perfect) level of reliability. (See chapter 9.) But only humans seem to have completely broken the 3-barrier when it comes to number. The only limit on how big a collection we can count is how much time we have available. Once we have learned the

counting trick as small children, there is no limit to how far we can count if we have the time.

Closely connected to counting is our use of arbitrary symbols to denote numbers and to manipulate numbers by the manipulation of those symbols. These two human attributes enable us to take the first step from an innate number sense to the vast and powerful world of mathematics.

The first thing to realize about counting is that it is not the same as simply saying how many members are in a collection. The number of members a collection has is just a *fact* about that collection. Counting the members of a collection, on the other hand, is a *process* that involves ordering the collection in some fashion, and then going through the collection in that order, counting off the members one by one.

We are so used to counting as a means of answering the question "How many?" that we forget we had to learn that counting tells you "how many." Very young children see counting and number as quite unconnected. Ask a three-year-old boy to count up his toys and he will perform flawlessly: "One, two, three, four, five, six, seven." He may well point to each toy in turn as he counts. But now ask him how many toys he has, and chances are he will tell you the first number that pops into his head. Or ask the average three-year-old girl to give you three toys, and she will probably give you however many toys she happens to collect together. And yet, if prompted, she will gaily recite the counting numbers "one, two, three, four, five, . . ."

Around four years of age, children realize that counting provides a means to discover "how many." Part of that realization is the recognition that, when you are counting the members of a collection, the order in which you count doesn't matter. The number you finally reach is always the same. From that moment on, counting collections of any size is just a matter of knowing how to use the number language: start with the initial sequence of basic number words "one" to "ten," then count "eleven, twelve," then use the "-teen" suffix up to "twenty," then start to use the "twenty-" prefix, and so on.

A reminder that counting is a skill we acquire, and not an ability we are born with, comes from studies of so-called primitive societies that do not have counting. (More precisely, they do not have counting beyond two, which arguably means they don't really have counting at all.) For example, when a member of the Vedda tribe of Sri Lanka wants to count coconuts, he collects a heap of sticks and assigns one to each coconut. Each time he adds a new stick, he says, "That is one." But if asked to say how many coconuts he possesses, he simply points to the pile of sticks and says, "That many." The tribesman thus has a type of counting system (or perhaps more precisely, a system for representing quantity), but it doesn't use numbers.

Or take the Warlpiris, an Aboriginal tribe in Australia. Their native language permits them to count up to two, after which everything is simply "many." The fact that the members of those tribes have no difficulty learning how to

count in English shows that it is not that they are unable to count. Rather, their native language harks back to a time when counting simply went: "one, two, many." (Other "primitive" peoples count "one, two, *three,* many." But never "one, two, three, *four,* many." The cut-off point for a universal number sense is three.)

How did our ancestors first develop the idea of counting, as opposed to estimating using their innate number sense? Well, they probably began the way young children do today, by using their fingers. As any parent or elementary school teacher knows, children learning arithmetic spontaneously use their fingers. Indeed, so strong is the urge for children to count on their fingers that if a parent or teacher tries to stop a child from using finger counting, insisting that the child do it the "right" way or the "grown-up" way, the child will simply use finger counting surreptitiously. And as for the notion that dispensing with fingers is the "adult" way, we all know that many adults use their fingers when doing arithmetic.

Certainly, some evidence that counting began as finger-enumeration is provided by the fact that our number system uses 10 as the base. Because we have ten fingers, if we use our fingers to count, we run out of counters when we reach ten, and so we have to find some way of recording that fact (perhaps by moving a pebble with our foot) and then starting again with our fingers. In other words, finger (and thumb) arithmetic is base-10 arithmetic, where we carry when we reach ten.

Further evidence in favor of the hypothesis that arithmetic began with finger manipulation is the fact that our word *digit,* meaning both a finger or toe and one of the basic numerals, derives from the Latin *digitus* (finger, toe).

Admittedly, neither piece of evidence is overwhelming on its own. But they are very suggestive when you combine them with some recent experimental evidence from neuroscience.

By using various techniques, scientists can measure the level of activity in different parts of the brain while it is engaged in a particular task. For instance, the frontal lobe is where most of the brain activity takes place when a person is using language. In a sense, the frontal lobe is the brain's language center.

Laboratory studies have shown that when a person is performing arithmetic, the most intense brain activity is in the left parietal lobe, the part of the brain that lies behind the frontal lobe. As it happens, similar studies have shown that the left parietal lobe is also the region that controls the fingers. (It requires a considerable amount of brain power to provide the versatility and coordination of our fingers, far more than any other parts of our bodies. Hence, a large part of the brain is devoted to that task.)

It is not a coincidence that the part of the brain we use for counting is the same part that controls finger movement. I believe it is a consequence of the fact that counting began (with our early ancestors) as finger-enumeration,

and that, over time, the human brain acquired the ability to "disconnect" the fingers and perform the counting without physically manipulating them.

In addition to the evidence from the neuroscience laboratories, clinical psychologists have also found a connection between finger control and numerical ability. Patients who sustain damage to the left parietal lobe often exhibit an unusual condition known as Gerstmann's syndrome, in which sufferers lack awareness of their individual fingers. For instance, if you were to touch a patient's finger, he or she would be unable to tell you which finger was being touched. Sufferers also are typically unable to distinguish left from right. More interestingly, from our point of view, people with Gerstmann's syndrome invariably have difficulty coping with numbers.

If our early *Homo sapiens* ancestors' first entry to the world of numbers, perhaps fifty or a hundred thousand years ago, was by way of their fingers, then the large region of the brain that controls the fingers would be the one in which their descendants' more abstract mental arithmetic would be located. Very likely, our present-day, strictly mental number sense is an abstraction from the physical finger manipulations of those early ancestors. Mental arithmetic may be, in essence, "off-line" finger manipulation, which became possible when our ancestors' brains acquired the ability to disconnect the brain processes associated with finger manipulation from the muscles that control the fingers.

Symbols of a Numerical Mind

Using our fingers to count indicates that we have a conception of numerosity, but it does not necessarily imply that we have a concept of *number,* which is purely abstract. If I say, "That jar contains five pennies," my statement is about the jar and its contents, not about numbers. The word "five" is an adjective modifying pennies. On the other hand, if I say "Think of the number five," I am using the word "five" as a noun. As such, it refers to a particular object. What object? *The number five.* The number five is not a concrete object like a chair, but an abstract object. We can't touch it or smell it. But we can think about it, and we can use it.

Those abstract objects we call numbers are the key to modern mathematics. They provide the transition from subconscious, innate mathematics—a capacity we share with many other creatures—to consciously created symbolic mathematics, which is almost unique to humans. How, and when, did we acquire them?

As long as thirty thousand years ago, our ancestors cut notches in wood and bone to keep track of (we believe) the passage of the seasons or the phases of the moon, and possibly other things as well. That was very definitely a process of counting, but it did not involve abstract numbers.

The best current evidence we have for the introduction of *abstract* counting numbers (1, 2, 3, and so on), as opposed to markings, was discovered by the University of

Texas archaeologist Denise Schmandt-Besserat in the 1970s and 1980s. At that time, Schmandt-Besserat was investigating archaeological sites in the Middle East, where the highly advanced Sumerian society flourished from about 3300 to 2000 B.C.

Wherever Schmandt-Besserat dug, she kept finding small clay tokens of different shapes, including spheres, disks, cones, tetrahedra, ovoids, cylinders, triangles, and rectangles. The older ones were more simple, later ones often quite intricate. At first she was puzzled by her findings. But gradually, as she and other archaeologists slowly pieced together a coherent picture of Sumerian civilization, it became clear that the tokens were physical counters used in commerce. Each shape represented a certain number or quantity of a particular item: metal, a jar of oil, a loaf of bread, an ox, a sheep, a garment, and so forth. (See figure 11.1)

Figure 11.1. These small clay shapes, found by Denise Schmandt-Besserat in the Middle East, were used by the Sumerians to count possessions between about 3300 and 2000 B.C.

As far as we know, this was the earliest organized form of counting (and accounting). Notice that there were still no abstract numbers. Because the clay tokens were used for counting, we could view them as a very concrete kind of "numbers." As such, they were to be the first step toward the abstract numbers we use today.

A Sumerian businessman or trader would keep all his clay tokens in one place as a record of his financial worth. Typically, he would place his pile of tokens onto a sheet of wet clay, which he then folded into a pouch and sealed. This method was certainly secure. Once the clay had set hard, there was no danger of losing track of his worth. The obvious problem arose whenever it came time to trade. The Sumerian had to break open the pouch in order to update his records, by adding or removing tokens. Even worse, he had to break open the pouch whenever he wanted simply to check his current balance.

To get around the frustration of constantly having to break the pouch and make a new one, the more enterprising Sumerians adopted the habit of pressing the clay tokens into the surface of the wet clay before sealing it up into a pouch. In this way, they left a record on the outer surface of the pouch of the tokens that were locked inside it. This meant that the Sumerian no longer had to break open the pouch simply to check his balance. All he had to do was examine the markings on the outside of the pouch.

And so things remained until (we presume) a particularly astute Sumerian realized that a further simplification

was possible. The way things were, the tokens in the pouch represented a certain quantity of goods. In turn, those tokens were represented by the markings on the pouch, caused by the tokens' being pressed into the wet clay before it set hard. But that meant you didn't need the tokens themselves! The crucial information lay in the markings on the outside of the pouch. You could do without the tokens altogether, and rely simply on markings on the clay. Then, of course, the clay no longer had to be fashioned into a closed pouch. It could be left in the form of a flat sheet, an example of which is shown in figure 11.2.

Figure 11.2. The beginnings of writing. When the Sumerians stopped storing clay tokens, and instead made markings on a wet clay tablet, they effectively invented abstract numbers, and took the first key step toward written language.

At that moment we have the origins of two of the most basic underpinnings of modern society. First, it marked the beginning of symbolic language. By "symbolic" I mean the use of standardized but essentially arbitrary symbols to stand for ideas, as opposed to recognizable drawings or pictures. According to Schmandt-Besserat, the Sumerian use of symbols to denote numbers came before the introduction of written language, where symbols are used to denote words. If this is true, the fact that the driving force behind the introduction of written symbols was numbers rather than words is yet another sign of just how fundamental numbers are to us.

The second fundamental change brought about by the Sumerian abandonment of the clay tokens in the envelopes is that it signified, to all intents and purposes, the birth of abstract numbers. For when the clay tokens were discarded, they left behind conceptual ghosts: abstract numbers–the things the symbols denote and which in turn represent the numerosity of collections of objects in the world.

Nowadays, numbers and the symbols that denote them are so much a part of the fabric of our lives that we rarely give them any thought. We might think about the steps of a computation or a problem in arithmetic, but not the numbers themselves. And yet they are one of humankind's most profound and powerful inventions, and they pervade almost every aspect of modern life.

The very abstractness of numbers means we have to

approach them, and manipulate them, through language. Unlike concrete objects such as cats, chairs, or other people, which we can think about without the need for words or other symbols to refer to them, numbers are closely tied to the symbols that denote them.

Some mathematicians may object to that last observation, and as a mathematician myself I understand their objection. There is a sense in which mathematicians, and possibly other people as well, can develop an ability to think about numbers separately from the symbols that denote them. But the link can never be completely severed. Dramatic evidence of this fact that was provided by two Israeli researchers, Avishai Henik and Joseph Tzelgov, in the early 1980s. They showed subjects pairs of digits in different-sized fonts on a computer screen, and measured the time it took the subject to decide which symbol was printed in the larger font.

On the face of it, this task has nothing at all to do with what number the digit denoted. The task involved purely the size of the actual symbol. Nevertheless, subjects took longer to respond when the size of the fonts conflicted with the relationship between the sizes of the numbers than when the two agreed. For example, it took longer to decide that the symbol 3 is larger than the symbol 8 than to decide that the symbol 8 is larger than the symbol 3. Subjects were unable to forget the fact that the number 8 is larger than the number 3. We are, it appears, unable to separate number symbols from the quantity they denote. (In contrast, people have much less difficulty choosing the

larger font when presented with pairs of spelled-out words for numbers, such as **Three** and Nine.)

Because we are so familiar with our number symbols and because many of us have been so taught in school, we have a tendency to identify a sequence of numerals such as 349 (i.e., a string of three symbols) with the number it denotes–to think of "349" as the number. This can lead us to overlook the fact that our number system constitutes a language. It is a language for naming numbers. In fact, it's the closest thing the world has to a genuinely international language. Although people in different parts of the world speak different languages, and in some parts of the world use different alphabets to write words, everyone writes numbers the same way, using the ten Arabic numerals 0, 1, 2, 3, 4, 5, 6, 7, 8, 9.

It is really quite remarkable that, by using just those ten numerals (or digits), we can represent any positive whole number. The idea that makes it work is so familiar to us that we rarely stop and reflect that it is an extremely clever device. Namely, we use the digits to form numerical "words" that name numbers, just as we put letters together to create words that name various objects or actions in the world.

The number denoted by a particular digit in any numerical "word" we write down depends on its position in that word. Thus, in the number

1492

the first digit 1 (in the thousands column) denotes the number *one*-thousand, the second digit 4 (in the hundreds column) denotes *four*-hundred, the third digit 9 (in the tens column) denotes *nine*-tens (or ninety), and the last digit 2 (in the units column) denotes the number *two.* Thus, the entire "word" 1492 denotes the number

one-thousand and *four*-hundred and *nine*-ty *two.*

This number system was developed in India and reached essentially its present form in the sixth century. It was introduced into the west by Arabic traders and scholars in the seventh century, and as a result is generally referred to as the "Hindu-Arabic number system," or more simply as the "Arabic system." It is one of the most successful conceptual inventions of all time.

Once the Arabic system was available for representing any positive whole number, it was easy to extend it to represent fractional and negative quantities. The introduction of the decimal point or the fraction bar allows us to represent any fractional quantity (3.1415 or $^{31}/_{50}$, for example). Introduction of the minus sign " - " extends the range to all negative quantities, whole or fractional. (Negative numbers were first used by sixth-century Indian mathematicians, who denoted a negative quantity by drawing a circle round the number; European mathematicians did not fully accept the idea of having negative numbers until the early eighteenth century.)

You get some idea of just how efficient the Arabic number system is when you think for a moment about one of the number systems that preceded it: Roman numerals, occasionally still used today in a largely ceremonial way.

The date 1492 in Roman numerals would be expressed

MCDXCII

M = one thousand + CD (one hundred [C] less than five hundred [D]) + XC (ten [X] less than one hundred [C]) + II (two [I and I]).

For the Romans, with their cumbersome number notation of I's, V's, X's, etc. even the simplest of arithmetic sums was difficult to perform symbolically. (Try a few simple additions and multiplications and see for yourself.) Moreover, the Romans had no way to represent fractional or negative quantities.

The Romans derived their system from the ancient Greeks. For all their prowess in abstract mathematics (particularly geometry), in their everyday lives the ancient Greeks used a very simple, cumbersome system for representing numbers. The starting point for the Greek number system is a system many of us use today when counting collections, such as the number of people in a tour group. We make individual tally marks on a piece of paper, grouping the tallies in fives by drawing a diagonal line

through the previous four tallies when the next (fifth) item in counted. For example, the string of symbols

卌 卌 卌 III

denotes 18 objects (5 + 5 + 5 + 3).

The Greeks likewise used vertical tally marks, but grouped in multiples of five, ten, and one hundred, using the first letter of the (Greek) words for each group, and writing down clusters of such symbols from left to right. For example, the Greeks would write the number 428 as

HHHHDDPIII

That is: 4 × H (*Hekaton,* or hundred) plus 2 × D (*Deka,* or ten) plus 1 × P (*Pente,* or five) plus 3 (units).

The Arabic number system was a tremendous advance over the systems that preceded it. Not simply because it makes computation much easier, but also because, with the Arabic system, the number "words" can be read aloud, and moreover, the spoken version reflects the numeric structure in terms of units, multiples of ten, multiples of a hundred, and so forth. For example, the Arabic number-word 5823 can be read aloud as the English language number phrase "five thousand eight hundred and twenty-three." (Below we shall consider the arithmetical consequences of the variations in reading Arabic numbers in different languages, such as Japanese or Chinese.)

Another strength of the Arabic number system is that it *is* a language. Consequently, it allows humans, who have an innate linguistic fluency, to use their ability with language in order handle numbers. Thus, whereas our intuitive number sense resides in the left parietal lobe, the "linguistic" representation of exact numbers is handled by the frontal lobe (the language center), as we shall see in more detail below.

Although the use of the symbols 1, 2, 3, 4, 5, 6, 7, 8, 9 for the digits is now universal, in the past there have been others, including cuneiform, Etruscan, Mayan, ancient Chinese, ancient Indian, and the Roman system mentioned already. The Chinese still use a modern variant of their ancient system in addition to Arabic notation, and of course Western societies still use Roman numerals for some specialized purposes.

In the light of our earlier discussions about our number sense and the special nature of the first three counting numbers 1, 2, 3, it is of interest to note that in all the number representation systems ever used, the first three numbers are denoted the same way: 1 is denoted by a single stroke or dot, 2 by placing two such symbols side by side, and 3 by placing three such symbols side by side. In the case of the Roman system, for example, the first three numerals are I, II, III. In Mayan notation, dots are used: •, ••, •••. The different systems started to differ from the fourth numeral onward.

It may appear that the Arabic system does not follow

that pattern, but it does. The ancient Indian system used horizontal bars, like this: −, =, ≡. Our present numerals arose when people started to write these three symbols without taking the pen from the paper, giving this kind of pattern: –, Z, ∍. At some stage, the first symbol became vertical, as in the Roman system, rather than horizontal. When printing was invented, the numerals were given the stylized versions we use today: 1, 2, 3.

The Arabic number system, including the rules (the "grammar") by which the digits are put together to form number "words," is based on the number ten. There is no mystery about this choice of number *base,* as it is called. As noted earlier, one of the earliest and most obvious ways people found to count was to use their fingers and thumbs–their digits.

Incidentally, the original idea of denoting numbers by having a small collection of basic symbols and stringing them together to form number "words" is due to the Babylonians, around 2000 B.C. Because it was built on the base 60, the Babylonian system itself was cumbersome to use, and thus did not gain widespread acceptance, although we still use it in our measurement of time (60 minutes make one hour, 60 seconds make one minute).

Other than being the natural base for counting on our hands, there is nothing indispensable about the base 10, and other bases have been–and are–used for various purposes. In particular, present-day computers use base 2 (binary) arithmetic, which is the most convenient system

for digital electronic devices. Our system for telling the time uses base 12 (or base 24 for some purposes, such as railway timetables).

As mentioned above, one of the most powerful aspects of Arabic notation is that arithmetic can be performed by means of fairly straightforward (and easily learned) formal manipulations of the symbols. When performing addition, for example, we write all the numbers one beneath the other, aligned in columns starting from the right, and then proceed to add the digits in each column from right to left. Whenever a sum in a column reaches 10, we put a zero in that column and carry a 1 to the next column to the left. This procedure can be carried out in an automatic fashion. The steps performed do not depend on what the actual numbers are. In particular, we can design machines to do it for us. The same is true for the other basic arithmetical operations, subtraction, multiplication, and division. For each operation, there is a standard procedure that always works, regardless of what the actual numbers are.

Arabic notation renders basic arithmetic such a mechanical process that, in the days before the widespread availability of cheap, hand-held calculators, the necessity of learning how to calculate made the elementary arithmetic classes in schools one of the least popular for most pupils. It is a great pity that for so many years our teaching methods have been such that, for many individuals, possibly the majority, one of humankind's greatest conceptual inventions is passed over without recognition, obscured

by the trivia of symbolic manipulation. The Arabic number system is an incredible human invention, concise and easily learned. It allows us to represent numbers of unlimited magnitude—numbers that can be applied to collections and measurements of all kinds. Moreover, and this is surely its greatest strength, it reduces computation with numbers to the routine manipulation of symbols on a page (or electrical pulses inside a computer). In fact, as far as I can see, the Arabic number system has just one drawback: It makes it very hard to learn our multiplication tables.

WHY YOU ARE NOT SURE WHAT 8 × 7 IS

As we have observed, our brains appear to handle number symbols differently from number words. Number symbols are closely bound up with the actual numbers (the points on our mental number line), whereas number words are "just names" for numbers. This hypothesis is born out by studies of people with pathological brain disorders.

For instance, there are people who are unable to read words, but who can read aloud single or multi-digit numbers presented to them using numerals. Conversely, there are some individuals who can read words, including number words and word expressions for multi-digit numbers, but are unable to read aloud a number of two or more digits presented to them in numerals.

The psychologist Brian Butterworth has reported an

extreme case of a woman named Donna who had surgery to the left frontal lobe of her brain. Although she can read and write single- or multi-digit numbers in numerals, not only can she not read or write words, she can only name about half the letters of the alphabet. Despite her inability to write even her own name—the result is an illegible scribble—when presented with a standard arithmetic test (where the questions are presented in purely numeric form) she does just fine, writing her numerals neatly, in columns, and invariably getting the right answer.[24]

The fact that we access numbers through language is the key to understanding why it is that many of us have difficulty in learning the multiplication tables. By rights, learning the multiplication table should be easy. There are, after all, only a few facts involved. If you had to learn the product of each number from 1 to 10 with each number from 1 to 10, there would be 100 separate facts. That's a tiny number when you consider that by the age of six, a typical American child will have learned between 13,000 and 15,000 words that he or she can recognize in context and know the correct meaning of. But there are far fewer than 100 multiplication facts you have to commit to memory. For one thing, no one has to learn the one-times table or the ten-times table. Discounting those, the entire collection of multiplication tables amounts to only 64 separate facts (each one of 2, 3, 4, . . . , 9 multiplied by each one

24. Butterworth, 1999, pp. 197–199.

of 2, 3, 4, . . . , 9). Most people have little problem with the two-times table or the five-times table. Discounting those leaves just thirty-six single-digit multiplications where it takes some effort to commit them to memory. (Each of 3, 4, 6, 7, 8, 9 times each of 3, 4, 6, 7, 8, 9.) In fact, anyone who remembers that you can swap the order in multiplication (for example, 4×7 is the same as 7×4) can cut in half the total to be memorized, to eighteen. So, the total number of individual facts that have to be learned to master all the multiplication tables is only eighteen. So why do we find it so difficult to remember them?

The problem has to do with language. We learn and remember our multiplication tables in terms of linguistic patterns, much as we learn a poem. Most commonly, when at elementary school, we are required to recite the tables over and over again. What we are learning are not so much facts about numbers as word patterns in language. Although regular use of the tables might lead to the brain going beyond those word patterns and developing genuine "number patterns," those linguistic patterns remain dominant. Even today, fifty years after I "learned my tables," I still recall the product of any two single-digit numbers by reciting that part of the table in my head. I remember the sound of the number words spoken, not the numbers themselves. Indeed, as far as I am aware, the pattern I hear in my head is precisely the one I learned when I was seven (complete with a Yorkshire accent).

Just as language provides the means for learning the

multiplication tables, so too it provides the key to understanding why we have so much difficulty with some multiplications. Why, despite many hours of practice and repetition in school, do adults of average intelligence tend to make mistakes roughly ten percent of the time? And why, for some particularly problematic multiplications such as 8 × 7 or 9 × 7, can it take up to two seconds to produce an answer, with the error rate going up to 25 percent? (Is 8 × 7 equal to 54, 56, or 63? What about 9 × 7?—another notorious one.)

The problem lies not with a weakness of the human mind but with two of its greatest strengths. First, the mind is a truly superb pattern recognizer. Just how good the human mind is at discerning patterns is made clear to us when we see a face in a landscape, in a rock formation, in an abstract wallpaper pattern, or on the surface of the moon. The second great strength of the human mind is its powerful mechanism for pattern (or memory) association. As we have all experienced, our memory works by association: one thought leads to another. Someone mentions Germany, and that brings to mind our vacation there three years ago, which reminds us that we need to decide where to go next year . . . but the roof needs repairing so maybe we should forgo our vacation to pay for that—whoops, we forgot to pay the contractor's bill for mending that wall. And so it goes, from Germany to a contractor's bill in just four steps, with one thought leading to another in a chain that could go on forever if we let it.

These two features of the human mind make it very different from a digital computer. Despite an enormous investment in money, talent, and time over the past fifty years, attempts to develop computers that can make sense of a visual scene have largely failed. And it is only in very limited ways that computer databases can be designed to carry out pattern association. On the other hand, we have difficulty doing some things that computers do with ease. Remembering our multiplication tables is one of them. Computers are well suited to precise storage and retrieval of information and to exact calculation. A modern computer can perform billions of multiplications in a single second, getting each one right.

Because we remember our table linguistically, many of the different entries interfere with one another. Whereas a computer "sees" the three multiples $7 \times 8 = 56$, $6 \times 9 = 54$, and $8 \times 8 = 64$ as quite separate and distinct from each other, the human mind sees similarities between these three multiplications, particularly linguistic similarities in the rhythm the words make when we recite them out loud. For instance, when we see the pattern 7×8, it activates several patterns, among which are likely to be 48, 56, 54, 45, and 64.

Stanislas Dehaene illustrates this point cleverly in his book *The Number Sense* with the following example (page 127): Suppose you had to remember the following three names and addresses:

- Charlie David lives on Albert Bruno Avenue
- Charlie George lives on Bruno Albert Avenue
- George Ernie lives on Charlie Ernie Avenue

Remembering just these three facts looks like quite a challenge. That is because there are too many similarities, and as a result each entry interferes with all the others. But these are just disguised entries from the multiplication tables. Let the names Albert, Bruno, Charlie, David, Ernie, Fred, George stand for the digits 1, 2, 3, 4, 5, 6, 7, respectively, and replace the phrase "lives on" by the equals sign, and you get the three multiplications

- $3 \times 4 = 12$
- $3 \times 7 = 21$
- $7 \times 5 = 35$

It's the *pattern interference* that causes our problems. The phenomenon of pattern interference is also the reason why it takes longer to realize that $2 \times 3 = 5$ is false than is does to realize that $2 \times 3 = 7$ is wrong. The former equation is correct for addition ($2 + 3 = 5$), and so the pattern "2 and 3 make 5" is familiar to us. In contrast, we don't know any pattern of the form "2 and 3 makes 7."

We see this kind of pattern interference in the learning process of young children. By the age of seven, most children know by heart many additions of two digits. But as they start to learn their multiplication tables, the

time it takes them to answer a single digit addition sum increases, and they start to make errors such as $2 + 3 = 6$.

Another way that linguistic pattern similarities interfere with retrieval from the multiplication table is when we are asked for 5×6 and answer 56. Somehow, reading the 5 and the 6 brings to mind this incorrect answer. On the other hand, people do not make errors such as $2 \times 3 = 23$ or $3 \times 7 = 37$. Because the numbers 23 and 37 do not appear in any multiplication table, our associative memory does not bring them up in the context of multiplication. But 56 is in the table, so when our brain sees 5×6, the number 56 is activated.

To repeat, much of our difficulty with multiplication comes from one of the most powerful and useful features of the human mind: its associative memory, with its great facility for pattern recognition. Those mental capacities evolved over hundreds of thousands and millions of years to meet the demands of the lives of our early ancestors. Those demands did not include doing arithmetic, which is at most a few thousand years old. To do arithmetic, we have to make use of mental circuits that developed (i.e., were selected for during the course of evolution) for quite different reasons.

So great is the effort required to learn the multiplication tables (because of the interference effects) that people who learn a second language generally continue to perform arithmetic in their first language. No matter how fluent they become in their second language—and many people reach the stage of thinking entirely in whichever language they are conversing in–it is easier to slip back into their first language to calculate and then translate the result back than to try to relearn the multiplication table in their second language. This observation formed the basis of an ingenious experiment that Stanislas Dehaene and his colleagues performed in 1999 to confirm that we use our language faculty to perform arithmetic.

The hypothesis they set out to establish was this: That arithmetical tasks that require an exact answer depend on our linguistic faculty–in particular, they use the verbal representations of numbers–whereas tasks that involve estimation or require an approximate answer do not make use of the language faculty.

To test this hypothesis, the researchers assembled a group of English-Russian bilinguals and taught them some new two-digit addition facts in one of the two languages. The subjects were then tested in one of the two languages. For questions that required an exact answer, the subjects took longer to answer when the question was posed in the language other than the one they had been taught in than it did when the question was asked in the same language that had been used for the instruction. When the question asked for an approximate answer, however, the language of questioning made no difference to the time of response.

According to the experimenters, the extra time required to answer an exact question in the "other" language (about 1 second longer than the 2.5 to 4.5 seconds it

took to answer in the "same" language) arose because the subjects translated the question to the language in which the facts had been learned.

To see what parts of the brain the subjects used when they were answering the different types of question, the researchers monitored the subjects' brain activity throughout the testing process. When the subjects were answering questions that asked for approximate answers, the greatest brain activity was in the two parietal lobes, the regions that house the number sense and support spatial reasoning. For questions requiring an exact answer, however, there was far more activity in the frontal lobe, where speech is controlled.

Altogether, the result was quite convincing: The ability of humans to extend the intuitive number sense to a capacity to perform exact arithmetic seems to depend on using our language faculty. But if that is the case, wouldn't we expect to see differences in arithmetical ability from one country to another? For, if the words used for numbers are significantly different, presumably this would reflect in how well or how easily people learn their tables. This is indeed what happens, as we discover next.

THE SOUND OF NUMBERS

Every few years, U.S. newspapers report that American schoolchildren have once again scored poorly in yet

another international comparison of mathematical ability. Although there is never any shortage of knee-jerk reactions to such news, it is extremely difficult to draw definite conclusions from cross-national and cross-cultural comparisons. Many factors are involved, and even if there is a real problem, simplistic solutions are unlikely to have much effect. Education is not a simple matter.

In such comparisons, Chinese and Japanese children often appear to outperform American children, as well as children from England and much of Western Europe, who tend to score roughly the same as those in the United States. Given the similarities in culture between the United States, England, and Western Europe, and the differences between the West and the cultures of China and Japan, it is reasonable to suppose that cultural factors, including the differences between the school systems, contribute to differences in mathematical performance. But so too does language. Learning to count and do arithmetic is easier for Chinese and Japanese children.

Part of the reason is that their number words are much shorter and simpler–generally a single, short syllable such as the Chinese *si* for 4 and *qi* for 7. This makes them much easier to recite, either aloud or internally, and that in turn makes them easier to learn. Not only are the words for the individual digits shorter in Chinese, but their grammatical rules for building up other number words are also much easier than in English. For example, the Chinese

rule for making number words for numbers past ten is simple: 11 is *ten one,* 12 is *ten two,* 13 is *ten three,* and so on, up to *two ten* for 20, *two ten one* for 21, *two ten two* for 22, etc. Think how much more complicated is the English system. (French and German speakers will know that it's even worse in those languages, with their tongue-twisting *quatre-vingt-dix-sept* for 97 and *vierundfünfzig* for 54.) A recent study by Kevin Miller showed that differences in language cause American children to lag a whole year behind their Chinese counterparts in terms of learning to count. By age four, Chinese children can generally count up to 40. American children of the same age can barely get to 15, and it takes them another year to get to 40. How do we know the difference is due to language? Simple. The children in the two countries show no age difference in their ability to count from 1 to 12. It's when they get beyond the number 12 that the differences start to appear, when the American children start to encounter the various special rules for forming number words. Chinese children, meanwhile, do not have to learn any new rules. They simply keep applying the same ones that worked for 1 to 12. (When American children try to apply the same rules, the teacher corrects them. "No, you can't say twenty-nine, twenty-ten, twenty-eleven. You have to say twenty-nine, *thirty,* thirty-one.")

In addition to the Chinese number-word system being easier to learn, it also makes elementary arithmetic easier, because the language rules closely follow the base-10 structure of the Arabic system. A Chinese pupil will *see*

from the linguistic structure that the number "two ten five" (i.e., 25) consists of two 10s and one 5. An American pupil has to *remember* that "twenty" represents two 10s, and hence that "twenty-five" represents two 10s and one 5.

Thus, when it comes to learning how to count and do elementary arithmetic, the language we speak can affect our performance. It does so because we rely on our linguistic ability in order to handle numbers. In consequence, patterns of language can help, or hinder, our attempts to learn how to count and perform certain arithmetic tasks.

One area where patterns of language definitely hinder mastery of arithmetic–and in this case it affects children of all nationalities–is in learning how to add fractions. For example, the following incorrect addition illustrates a common error in adding fractions:

$$\frac{1}{2} + \frac{3}{5} = \frac{4}{7}$$

A person who makes this mistake sees the problem as two addition sums: first add the numerators 1 + 3 = 4 and then add the denominators 2 + 5 = 7. This is the most logical thing to do from a symbolic point of view.[25] It is incorrect because it makes no sense in terms of adding the

25. It is also correct arithmetically if you think (wrongly) that adding fractions is about combining proportions. If you have 2 people of whom 1 is a woman, and another 5 people of whom 3 are women, then altogether you have 7 people of whom 4 are women.

numbers the symbols represent. The symbolic manipulations you have to perform to get the correct answer (that is, the symbolic manipulations that correspond to adding the actual fractional numbers represented by the symbol-words ½ and ⅗) are fairly complicated. What is more, those symbol-manipulation rules only make sense if you think of the numbers the symbols represent. Purely as rules for manipulating symbols, they make no sense at all.

I am sure that it is because of cases such as this that many children come to see mathematics as "illogical" and "full of rules that make no sense." They think of mathematics as a collection of rules for *doing things with symbols*. Some symbolic rules seem sensible; others appear quite arbitrary. The only way to avoid this misconception is for teachers to ensure that their pupils understand what the symbols represent. Often this is not done. Nevertheless, some children do learn how to add fractions correctly. How is this possible?

Since the human mind is an excellent pattern recognizer with tremendous adaptive powers, with enough training it can adapt to be able to perform almost any symbolic procedure in an essentially "mindless" fashion. Thus, it is possible to train a human mind to carry out a procedure such as manipulating the symbols required to add to fractions correctly:

> Start out by multiplying the two denominators. That will give you the denominator in the answer.

Then multiply the numerator of the first fraction by the denominator of the second, and the numerator of the second fraction by the denominator of the first, and add those two results. That gives you the numerator in the answer. Then see if there are any numbers that divide both the numerator and the denominator in your answer, and if there are, divide both numerator and denominator by that number. Repeat this double division until you can't find any such common divisors. What's left is your final answer.

For example, to add $3/7$ to $4/9$, you proceed like this:

$$\frac{3}{7}+\frac{4}{9}=\frac{\text{something}}{7\times 9}=\frac{(3\times 9)+(4\times 7)}{7\times 9}=\frac{27+28}{63}=\frac{55}{63}$$

Written out using algebra, the rule gives the formula

$$\frac{a}{b}+\frac{c}{d}=\frac{\text{something}}{b\times d}=\frac{(a\times d)+(c\times b)}{b\times d}$$

However you write it, it's a complicated-looking procedure. From a symbolic (i.e., linguistic) point of view, it makes no sense at all. Symbolically, the most "sensible" rule would be

$$\frac{a}{b}+\frac{c}{d}=\frac{a+c}{b+d}$$

which is numerically wrong. However, despite the complexity, with sufficient practice most people can learn to follow the correct rule. Evolution has equipped us with a brain that can learn to perform particular sequences of actions. But unless someone shows you *why* each step is being performed—that is, shows you what is going on in terms of the *numbers* represented by the symbols—the whole thing is just so much mumbo jumbo. Of course, many children memorize how to perform the mumbo jumbo and manage to get an A for the course. But because they don't understand what they are doing, the moment the final math exam is over, they forget the complicated rules they learned, and they leave school unable to add fractions. But if they understood what was going on, they would never forget the procedure.

Another example of the problems that can arise when we blindly apply a symbolic rule without linking the symbols to the numbers they represent is provided by those word puzzles that you find in "brain teaser" books. For example:

- A farmer has 12 cows. All but 5 die. How many cows remain?
- Tony has 5 balls, which is 3 fewer than Sally. How many balls does Sally have?

Many intelligent people will get one or both of them wrong. The reason is a confusion of two patterns, one in

everyday language, the other in number symbols. Given the numbers 12 and 5 in the first problem, together with the question "How many remain?" there is a strong temptation to see the problem as asking you to perform the subtraction 12 - 5. Thus, many people give the answer 7. The correct answer is 5. But to get that correct answer, you have to think what the problem is actually saying. Blindly rushing to the symbolic manipulation stage *sometimes* works, but not on this occasion.

In the second problem, you see the numbers 5 and 3 together with the word "fewer," and the temptation is to perform the subtraction 5 - 3, giving 2 as the answer. Again, a hasty leap to the symbolic manipulation has led you to the wrong answer. When you stop and think what the question is saying, you realize you should *add* 3 to 5. The correct answer is that Sally has 5 + 3 = 8 balls. Tony has 3 fewer than Sally's 8, which means that Tony has 5 balls, as the question stated.

Again, our dependency on language ability in order to handle numbers, so useful in many respects, comes at a price. Unless we exert considerable effort to go beyond the symbolic and linguistic patterns to the numbers the symbols denote, the brain's natural facility with language and patterns of language can get in the way of our doing arithmetic.

While on the topic of patterns of language, let me tell you about a friend of mine who uses linguistic patterns to dramatically good effect. Arthur Benjamin is a

mathematician who can perform amazing feats of mental arithmetic—so much so that he has a successful second career as a stage performer, dazzling the audience by performing difficult calculations with numbers they shout out to him on the spot.

A few years ago I attended a luncheon at which Benjamin was giving a demonstration of his arithmetical skill. Just before he was due to begin, he asked the organizers to have the air conditioning turned off. While we were waiting for that to be done, Benjamin explained that the hum from the system would interfere with his calculations. "I recite the numbers in my head to store them during the calculation," he said. "I have to be able to hear them, otherwise I forget them. Certain noises get in the way." In order words, one of Benjamin's "secrets" as a human calculator is his highly efficient use of linguistic patterns—the sounds of the numbers as they echo in his mind.

Although few of us can match Benjamin in terms of calculating square roots of six-digit numbers, like him we depend on patterns of language in order to handle numbers. One of the secrets to being "good with numbers" is learning how to use our linguistic abilities to our advantage, rather than have them interfere with our attempts to do arithmetic, as is so often the case.

Another lesson we can learn from the lightning calculators is that much of their success stems from numbers having meaning for them. For example, to most of us, even

those of us who are comfortable with numbers, when we see a number such as 587, it's just a number. But to a calculating wizard, that number-word 587 may well have meaning–it may conjure up a mental image–just as the English word "cat" has meaning for us and conjures up an image in our minds.

There are, of course, a few numbers that have meaning for all of us. Americans see meaning in the numbers 1492 (Columbus's discovery of America) and 1776 (the signing of the Declaration of Independence), the English see meaning in the number 1066 (the Battle of Hastings), and anyone with a technical education sees meaning in the number 314159 (the start of the decimal representation of the mathematical constant π). Other numbers that have meaning for us, and which we therefore remember, are our Social Security number or other ID number, birthdate, and telephone number.

For a calculating wizard, however, many numbers have meaning. For the most part, that meaning lies not in the everyday world of dates, ID cards, and telephone numbers, but in the world of mathematics itself. For instance, Wim Klein, a famous calculating wizard who in the days before electronic computers once held a professional position with the title "computer," observes, "Numbers are friends to me." Of the number 3,844, he says, "For you it's just a three and an eight and a four and a four, but I say, 'Hi, 62 squared!' "

Because numbers have meaning for Klein and other

calculating wizards, calculation is meaningful to them. Consequently, they are much better at it. In fact, given the right circumstances–namely, a context in which the numbers have meaning–it is possible that every one of us could be a calculating wizard.

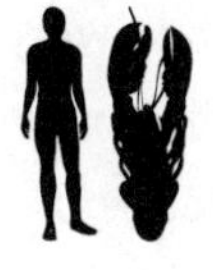

12.

The Trouble with Meaningless Math

The numerical dexterity of the young Brazilian street vendors (chapter 10) shows that they had developed considerable familiarity with numbers. To simplify computations, they used properties of the specific numbers they were faced with. Their standard approach was to find a way to transform the problem into one involving numbers and arithmetic operations they could recognize and handle. Sometimes this involved "rounding" the given numbers to ones that were easier for them to work with, and then going back and adjusting the result to correct for the rounding. On other occasions, they would break the initial problem into two or more subproblems. None of their methods were taught in school; their street math was very different from school mathematics. Since the children in the study—and children in other studies carried out by Nunes and her colleagues and by other researchers—exhibited far greater mastery of street mathematics than of school mathematics, we have to wonder, why is there such a big difference? This is the question we turn to next. What are the factors that make street mathematics work when school mathematics does not?

In many ways this is the key question in our story. It is fascinating to learn about the amazing things that animals can do with their built-in, natural mathematical ability, and to marvel at how challenging it can be for human mathematicians, scientists, and engineers to achieve the same results. Ultimately, though, the aim is to make use of what we have learned. Do nature's own mathematicians or the street vendors of Recife have anything to offer us that might help us to improve the teaching and learning of mathematics?

An obvious key factor for the Brazilian street traders is that, when the children carried out computations at their stalls, both the numbers and the operations they performed on them had *meaning,* and the operations made sense. Indeed, the children were surrounded by physical meanings of the arithmetical procedures they performed.

In contrast to street mathematics, the essence of

school mathematics is that it is entirely *symbolic.* In performing a standard school procedure for addition, subtraction, multiplication, or division, you carry out the very same actions, in exactly the same order, regardless of what the actual numbers are or what they measure. This is the whole point. The methods taught in school are supposed to be universal. Learn them once and you can apply them in any circumstance, whatever specific numbers are involved.

In the hands of a person who can master the abstract, symbolic procedures taught in school, those procedures are extremely powerful. Indeed, they underlie all our science, technology, and modern medicine, and practically every other aspect of modern life. But that doesn't make them easy to learn or to apply.

The problem is that humans operate on meanings. In fact, the human brain evolved as a meaning-seeking device. We see, and seek, meaning anywhere and everywhere. A computer can be programmed to obediently follow rules for manipulating symbols, with no *understanding* of what those symbols mean, until we tell it to stop. But people do not function that way. With considerable effort, we can learn our multiplication tables and train ourselves to follow a small number of arithmetical procedures. Even then, meaning is the key. Mastery of school arithmetic involves the acquisition of some kind of meaning for the objects involved and the procedures performed on them. It is doubtful whether it is even

possible for the human brain to perform a meaningless operation.

Since the arithmetic procedures taught in school are designed to be universal–to apply in all cases, whatever the actual numbers–the first thing a student has to do in order to master those procedures is to learn to ignore temporarily any possible meanings in terms of actual numbers or real objects in the world. The second thing the student must do to achieve mastery is to construct a different, more abstract kind of meaning. But the majority of students never get that far. They find themselves struggling to remember and apply seemingly meaningless sequences of operations on insignificant symbols. As a result, the answers they come up with are generally meaningless as well.

Any school mathematics teacher can tell stories of students who have given answers that make absolutely no sense: negative numbers for areas or volumes, negative weights, fractional numbers of people, annual salaries less than weekly pay, and so forth. Recall for example, the Brazilian girl from the street market who, having correctly worked out in her head the price of twelve lemons at Cr$5 each, gave the answer 152 when asked to calculate 12×5 on the test. There was also the child vendor who correctly computed in her head the change from a Cr$500 bill for two coconuts costing Cr$40 each (a task involving the subtraction $500 - 80 = 420$), but then gave the answer 130 when presented with the addition $420 + 80$ on a written test. (Her method was to add 8 to 2 to give 10, carry the 1,

add it to 4 and 8, to give 13, and write down the final 0 in the units column.) Neither child would accept at her stall the ridiculous answers she produced in the classroom.

Another example of an individual consistently applying an incorrect procedure comes from the research of the educational psychologists Lauren Resnick and Wendy Ford. A young boy in an American school produced the following answers on a test of basic addition:

7	9	17	87	365	657	923	27,493
8	5	8	93	574	794	481	1,509
15	14	25	11	819	111	114	28,991

He gets the first three correct, but as soon as he is faced with additions involving pairs of numbers with two or more digits, things go drastically wrong. He proceeds from right to left, column by column, as he should. Moreover, he can correctly add together pairs of digits. But every time the addition in a column gives rise to a carry, he writes the carry beneath the line and then moves on to the next column to the left. Since he does this consistently, he is clearly following a specific procedure. Moreover, he applies that procedure correctly, getting the "correct" answer each time, according to the procedure. Presumably when first taught the standard method for addition, he was confused, and ended up mastering a corrupted version of the correct procedure.

This pupil was not unintelligent–the example shown

indicates that he can consistently and "correctly" apply a multi-step, abstract procedure. Thus, had he really understood the correct procedure in terms of what each individual step does, he would not have gone astray. It is only because he viewed the procedure as an arbitrary set of rules having no meaning that he ended up applying a procedure that was arithmetically absurd.

With multiplication, the boy made a similar error (writing down the carry) as well as another error: proceeding from right to left in a strict column-by-column basis as in addition. Thus, he produced the following solutions to multiplication problems:

$$\begin{array}{r} 68 \\ \times 46 \\ \hline 24 \end{array} \qquad \begin{array}{r} 734 \\ \times 37 \\ \hline 792 \end{array} \qquad \begin{array}{r} 543 \\ \times 206 \\ \hline 141 \end{array}$$

Apart from thinking that $4 \times 0 = 4$ in the last example (a mistake many people make when multiplying by zero—the correct rule is $4 \times 0 = 0$), every step is arithmetically correct according to the procedure he is following. But again, it is not the right procedure. An intelligent person could only follow such a procedure if it was divorced from any meaning in terms of manipulating numbers.

The problem of poor mastery of standard arithmetical procedures is not limited to schoolchildren. Nunes, Schliemann, and Carraher carried out another study in Brazil, this time of (adult) carpenters. The researchers compared

the abilities of experienced carpenters and beginning apprentices to calculate the amount of raw timber required to construct a wooden bed-frame of certain dimensions. The carpenters, who were largely uneducated, all did fine. The same was not true for the apprentices, each of whom had had between 4 and 9 years of daily mathematics instruction at school. Having had no experience in performing such calculations on the job, the apprentices made use of the only numerical tools they had: the arithmetical procedures they had been taught in school. As a result, they produced answers that were wildly inaccurate. One apprentice calculated the amount of wood required to construct one bed measuring 1.9 meters long by 0.9 meters wide to be a block measuring 16.38 meters long by 10.20 meters wide by 0.12 meters thick. He arrived at this answer by adding together all the lengths of all the individual pieces of wood required, all the individual widths, and all the individual thicknesses.

Nor is the production of nonsensical answers restricted to individuals of low intelligence or poor education. Indeed, as I know from many years of experience as a college mathematics professor, college students produce the same kinds of "obviously" wrong answers as well. Even college mathematics students sometimes produce nonsensical answers when applying the intentionally "meaningless" symbolic procedures of arithmetic or higher mathematics.

The fact is, when otherwise sensible and capable people

are faced with school mathematics, reason and common sense often fly out of the window. It's not that someone who gives a ridiculous answer to a math problem can't see how silly it is when it is pointed out. If they are asked to repeat the "same" calculation in a context or fashion with immediate practical significance, they generally get the right answer, or at least an answer that is plausible. And they do even better if presented with a "real life" problem that they can handle using "street math"–arithmetic methods they have developed themselves "on the job." For example, in the case of the Brazilian schoolchildren tested by Nunes and her colleagues, in performing additions, 30 percent of their written answers (using the school methods) were over 20 percent off the right answer, whereas only 4 percent of their oral answers were over 20 percent off. For subtraction, 61 percent of their written answers were over 20 percent off, compared with just 11 percent of their oral answers over 20 percent off.

Although many of the methods that people use when performing street mathematics are number-specific and unlike the standard methods taught in school, in some cases there is almost no *procedural* difference between a street method and its schoolroom counterpart. And yet the difference caused by the absence of meaning in school mathematics is significant.

For example, the world over, practically everyone can handle money fluently. In most countries, there are two units of currency, with one unit being equal to a hundred

of the others. In the United States the two units are dollars and cents, with 100 cents being equal to 1 dollar. Practically every American, from an early age, becomes fluent at handling money. They do not confuse dollars with cents, and they know that, say, 159 cents is the same as 1 dollar and 59 cents. And yet, procedurally, handling dollars and cents is no different from working with our Hindu-Arabic positional number system, where the position a digit occupies signifies its value and where the key to using the standard processes for addition, subtraction, and multiplication is to keep track of which position is occupied by which digit. True, school arithmetic questions can be more complex than simply totaling prices or working out change, but as the Brazilian children demonstrated, they can cope with considerable complexity in their mental arithmetic, so complexity is not the real issue. Rather, what makes the difference is that money means something but the number-symbols that are written down in school arithmetic do not.

In short, street mathematics is all about carrying out meaningful operations on meaningful objects, whereas school mathematics is about carrying out purely formal manipulations of symbols whose meaning, if any, is not represented in those symbols. For most people, $27.99 means something, but 27.99 does not—it's "just a number."

Just how successful a person is at mastering school mathematics is largely a matter of how much meaning they can construct for the symbols manipulated and the

operations performed on them. Doing school arithmetic, even long division, does not involve procedures any more difficult or complex than the numerical manipulations you can observe a poorly educated nine-year-old child carrying out at a stall on a street corner in Brazil. The only difference is the degree of meaning involved. Indeed, once the meaning is there, school mathematics is much easier. With school arithmetic, once you have mastered the four standard procedures for addition, subtraction, multiplication, and division, you don't have to learn anything else—you just keep on applying those four standard procedures, regardless of the actual numbers you are faced with. It's so routine, we can build machines to do it for us. In contrast, street mathematics requires a whole bag of tricks, and depends on being able to find ingenious simplifications or groupings that depend on the actual numbers involved. The problem many people have with school arithmetic is that they never get to the meaning stage; it remains forever an abstract game of formal symbols.

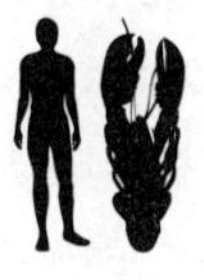

13.

Tapping into Our Math Instinct

I hope that by this point you have come to appreciate that there are two kinds of mathematics. There is the stuff most people think of when they hear the word "mathematics," namely the subject taught to children in school. I shall call this *abstract mathematics.* Then there is the kind of innate mathematics that I have described in the early part of this book, what I have called *natural mathematics.*

In fact, both abstract and natural math are just mathematics. The distinction lies in how the mathematics comes to get done. Abstract mathematics is symbolic and rule-based. To do abstract mathematics you have to *learn* what the symbols stand for and you have to *learn* how to follow the rules.[26] Natural mathematics arises, well, naturally. In the preceding chapters we have seen several different examples of natural mathematics, in both humans and other species.

Through the mechanism of evolution by natural selection, nature has crafted creatures that are purpose-built to perform by their very physical actions the natural mathematical computations of motion. Nature has equipped at least some species with visual systems that, by virtue of natural mathematical computations in the brain, enable them to see the world as three-dimensional. Nature also makes use of mathematics to provide certain animals with skin patterns that help ensure their survival in a hostile world, and has equipped many creatures with innate capacities (involving natural mathematics) that enable them to find their way from place to place and to catch prey.

Nature has also provided some species, among them pigeons, ravens, rats, lions, dolphins, monkeys, chimpanzees, and humans (to name some for which this capacity has been conclusively demonstrated), with another natural mathematical capacity: a sense of the size of a collection.

26. This does not mean that there is no place for creativity in formal mathematics. The rules simply establish the framework within which the mathematician must work. Also, the formulation of the rules in the first place is often a highly creative act.

In the case of human evolution, our ancestors also acquired another capacity: the ability to do abstract mathematics. Instead of relying solely on a small number of purpose-built, highly specialized, innate mathematical tricks (natural math) as other creatures do, we took advantage of that additional capacity to develop abstract mathematics, which provides a general-purpose toolbox for solving many different problems.

How did that capacity for abstract mathematics arise, and when? And how exactly are abstract and natural mathematics related? Before answering those questions, I should point out that, as the example of the young vendors in Recife showed, the same mathematics—in that case elementary arithmetic—can be both abstract and natural. They used natural mathematics when they carried out transactions in the market, and learned (or in most cases failed to learn) abstract mathematics in school.

How did we acquire the ability for abstract mathematical thinking?

One of the most puzzling aspects of the human capacity for abstract mathematical thinking is how our ancestors ever acquired it. Most abstract mathematics is at most two and a half to three thousand years old, depending on what you regard as the start of abstract mathematics. Numbers themselves are just ten thousand years old.

That's far too short a time frame for there to have been any major structural changes in the brain—evolution occurs over hundreds of thousands if not millions of years. When we do abstract mathematics, we have to be doing it with what is essentially an Iron Age brain. In other words, doing mathematics must involve taking mental capacities that our ancestors acquired for other purposes (more precisely, capacities that got into our gene pool because they proved advantageous for certain functions that were important to the survival of our early ancestors), and co-opting them for this new purpose. What are those capacities, when did our ancestors acquire them, what advantages did they confer, and what brought them together to yield a capacity for abstract mathematics?

These are the questions I answered in my book *The Math Gene*.[27] According to the thesis I advanced there, abstract mathematical thinking is an amalgamation of nine basic mental capacities that were acquired over a long time span of human evolutionary development. The complete story is more complex than I am hinting at here, but I can give a brief summary of my account as it pertains to arithmetic.[28]

In simple terms, the principal function of our brain is to ensure our survival, at least to the point where our

27. Basic Books, 2000.

28. *The Math Gene* addressed the issue of the evolutionary development of the ability to do all of mathematics, not just arithmetic.

offspring can get by on their own. The development of numerical skills (in a context) is at most a secondary feature. As we have observed, the innate number sense we are all born with is possessed by many other living species. Thus, it seems likely that such a sense confers a definite survival advantage from which many species have benefited. It is not hard to come up with plausible examples of such advantages. For example: knowing whether your group, tribe, or pack is outnumbered by a group of potential aggressors can help you decide whether to take flight or stay and defend your territory; finding your way back to your cave might require knowing how many hills or trees to pass before turning; and there is a considerable advantage to determining which tree bears the most fruit, and thus should be climbed first.

For a species that acquires language and starts to develop a complex societal structure, such as our *Homo sapiens* ancestors some two hundred thousand years ago, there are also clear advantages to being able to extend our innate number sense to handle larger collections with precision, which is what we achieve with counting. But since numbers and arithmetic are so recent, their use cannot have had any measurable effect on the early evolution of the human brain. Rather, numbers must have come about as a result of another evolutionary development. According to the development I outlined in *The Math Gene,* the key step that prepared the human brain to handle numbers was the acquisition of language about

one hundred thousand years ago. More generally, the acquisition of language was the final mental building block required to produce a brain that was capable of doing not only precise arithmetic but all of the deliberate, conscious, pencil-and-paper kind of mathematics I am calling abstract mathematics.

By language I don't simply mean using words, which possibly started to creep into use as long as two million years ago. Rather, language is the ability to assemble words into meaningful units (what we now call sentences) to express complex ideas. Many creatures have developed sophisticated communication systems, and in several cases (e.g. dolphins) it is not unreasonable to classify some of their communication signals as words. But only modern humans, *Homo sapiens*, have acquired language.

According to the argument presented in *The Math Gene*, the capacity for abstract mathematics resulted from a marriage of language (more precisely, the *capacity for* language) and the innate, instinctive mathematical abilities all humans have, many of which we share with other creatures. We can express this by means of a simple formula:

innate natural mathematical capacities + capacity for language → capacity for abstract mathematics

How do we do abstract mathematics?

Shortly after the publication of *The Math Gene,* the cognitive scientists George Lakoff and Rafael Nuñez published their book *Where Mathematics Comes From.*[29] Although written independently of mine, and in no way relying on my evolutionary account, by good fortune their book can be viewed as picking up exactly where mine left off. They describe in considerable detail how a brain that developed to handle the real (i.e., predominantly physical) world can think about mathematical abstractions. The key step is what they refer to as a formal (as opposed to a literary) metaphor. By this, they mean understanding something new and unfamiliar in terms of something familiar and already understood.

For example, you can understand positive counting numbers, and form a picture of them, as points arranged on a line starting at 0 and going from left to right, like this:

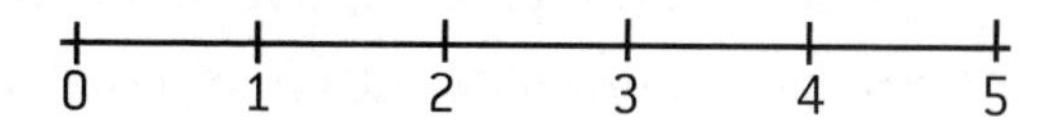

This creates an understanding of the counting numbers in terms of the familiar, everyday notion of a row of objects that can be examined one after the other. According to Lakoff and Nuñez, the brain learns to process unfamiliar (and perhaps abstract) concepts by co-opting existing

29. Basic Books, 2001.

"brain circuits" by way of formal metaphors. In particular, the points-on-a-line metaphor enables the brain to make use of its everyday ability to reason about objects in a line in order to process numbers. The metaphors used in this process do not have to be consciously created, and in fact in most cases they probably are not. It is more a matter of making use of mental capacities that arose for one purpose and making use of them for another.

Once you have the point-on-a-line metaphor for positive counting numbers, you can understand negative counting numbers as an entirely similar sequence going from right to left:

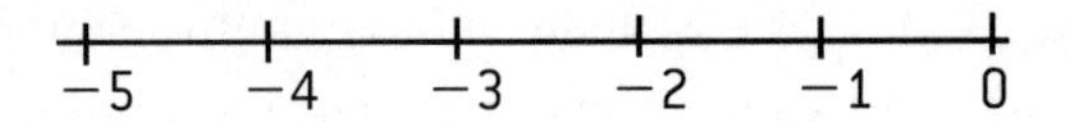

And so on.

The Lakoff and Nuñez theory is an attractive one, that builds upon the fact (dwelt on in the previous chapter) that our brains developed to process thoughts that have meaning for us, and that what causes us problems is trying to cope with abstractions, initially possibly meaningless to us, and something for which our evolutionary development did not prepare us.

According to Lakoff and Nuñez, we can form a simple picture of the growth of mathematical ability in a child as she grows up. Through play, the child first learns about shapes, collections, lengths, areas, volumes, points-on-a-line, rotations, and so forth. Using her ability to process

collections and a sense of the size of a collection, eventually the child forms the concept of number. In processing numbers, the child's brain co-opts various circuits that developed in order to deal with the physical world she lives in. With practice, with growing familiarity with the number concept, the combination of those individual circuits eventually functions as a whole. We may as well call that whole a "number circuit." At that point, it in turn may be co-opted to perform other functions. And so the entire process can repeat itself.

In their book, Lakoff and Nuñez trace out a long sequence of metaphors that starts with everyday mental processes about the physical world and, they claim, progresses through all of K-12 mathematics and on to university math. (The authors finish their account with the famous Euler equation $e^{i\pi} = -1$, but they claim that the chain can be taken as far as you want.)

If these authors are correct, then there is in principle no barrier to prevent people from mastering all the math they ever find they need. Each step forward involves essentially the same metaphor construction process. Crucially, during each metaphor construction step, the brain operates on and with concepts that have meaning–something the human brain evolved to do, and hence does well. The construction of the new metaphor amounts to a search for meaning of the new concept in terms of the old–and as we observed in the last chapter, searching for meaning is instinctive to the human brain.

The main limiting factors on this process are the length of time it takes to construct the appropriate metaphor(s) and how much practice with the new concept is required before the mind accepts it into its repertoire of familiar and understood concepts. The former can be greatly speeded up through teaching–indeed, in Lakoff and Nuñez's framework, teaching is essentially just the development by the student of suitable metaphors. And the latter is dependent on just how much time and effort is devoted to practice.

Since the metaphor chain is grounded in everyday thought processes, the entire process can be described as one of *abstracting and formalizing common sense.* Lakoff and Nuñez's basic thesis then amounts to a claim that all of mathematics is abstracted and formalized common sense.

I am not alone in suspecting that Lakoff and Nuñez overstate their claim, and that their argument breaks down about halfway through their book.[30] However, the breakdown, if one occurs, comes with some of the advanced mathematics taught in universities, which I and others believe requires a highly specialized form of thinking that cannot be viewed as formalized common

30. The investigation of what has come to be called mathematical cognition is very new, and there is as yet little by way of an established body of literature on the subject. But for some recent research findings to support the claim I am making here, see in particular the paper "Mathematical Thinking and Human Nature," by Uri Leron, ICME, 2004.

sense. (In fact, some advanced mathematics involves thought processes that are *counter* to common sense.) But that takes us beyond the scope of this book. When it comes to the kind of mathematics that most of us encounter in our everyday lives, everyone seems to agree with the Lakoff and Nuñez theory. In other words, as far as most people are concerned, abstract mathematics really is just formalized common sense. The question, then, is, can we find a way to tap in to our innate mathematical ability to improve our math?

As we have already seen, under certain circumstances the answer is a definite yes. For example, when ordinary people need to do arithmetic in a real-world context that matters to them, they can generally do so. If the requirement is fairly minimal, such as for the price-conscious shopper, they will find ways to perform the calculations they need with sufficient accuracy to meet their needs. In more pressured situations, where more is at stake, and where they have to perform many similar calculations day after day, they can achieve an impressive degree of numerical facility with almost total accuracy. What is of interest, however, is that in virtually all of those cases, although those people get the same answer that can be obtained using the arithmetical methods taught in school, that's not the way they do it. So what do you do if you have to improve your ability in traditional school math, say, to pass a test to get a new job?

How to improve your math skills

If you find yourself having to improve your ability at school math, there is a four-step approach that I would recommend.

The first step is to be aware that mathematical activity is a natural thing that occurs all the time in nature. (I hope that reading this book has convinced you of that fact.) Knowing that mathematics is natural should help overcome the fear that the subject so often evokes.

The second step is to approach abstract (i.e., school) math as merely a formalized version of your innate mathematical abilities–that is, as formalized common sense. In mathematics as in most other things in life, your approach can make all the difference in how well you do.

The third step is to recognize why the school methods were developed, what their advantages are, and what it is about them that makes them hard to learn. Knowing why something is done a certain way generally helps us to handle it.

Introducing the fourth step requires some preparation. In order for the general arithmetical (and other mathematical) procedures of abstract math to be universally applicable, which is their great strength, they have to be stripped of all context and taught in an abstract manner. But as we saw in the previous chapter, that is problematic for a brain that evolved to process things that arise in a particular context and have meaning. If the

human brain were some sort of universal, one-size-fits-all computational device that works best by applying the same generic tools to a variety of tasks, then teaching people the most universal methods should indeed be the most efficient approach. But all the evidence (masses of it) points the opposite way. The brain does not seem at all suited to acquiring general, universal skills and then applying them in particular circumstances. Rather, its strength seems to be its ability to solve problems when they arise in practice, developing on the job the skills and abilities needed. That includes the development of numerical or arithmetical skills, as we saw in the case of the Brazilian street vendors or the young bowling league scorer.

Of course, to those of us who can do abstract mathematics and can see how the very general procedures taught in the school math class can be used in many different circumstances, it might appear inefficient for people to keep "reinventing the wheel" each time they meet a new situation that involves arithmetic. But it's not at all inefficient. It's the natural way for the brain to do it. (Remember, no one had to struggle for years to teach the young bowling scorer how to perform the complicated calculations he used effortlessly in the bowling alley. But his teacher struggled for years without success to teach the boy elementary school arithmetic.)

Fortunately, if we do find a need to do better at abstract math, the highly adaptive human brain has a feature that

can come to our rescue. And this brings us to the fourth and final step of my four-step approach: Practice. Given sufficient repetition, our minds and/or our bodies can become skilled at performing practically any new task, be it swimming, riding a bicycle, typing, understanding and speaking a foreign language, or memorizing a poem. Your grandparents knew that instinctively. These days we can provide a scientific explanation that was not available in their day: the acquisition of such skills amounts to the selective development (i.e., creation and/or strengthening) of various neural pathways in the brain. In the case of learning abstract math, when the strength of those pathways starts to resemble the strength of the pathways associated with (and activated by) familiar, concrete objects in the real world, then the brain begins to experience the causes of the activation of those new pathways as being "real" or "concrete." In other words, for the human brain, familiarity breeds a sense of concreteness. And there we have the key to learning how to handle abstract entities: become sufficiently familiar with them for them to become (i.e., to seem) more concrete. No particular skill is required to do this. All it takes is sufficient repetition.

Of course, repetition of one particular task, or a particular set of tasks, can rapidly become tedious, be it learning to play the piano or learning how to add fractions. It would be nice if there were some other way, but there isn't. We humans have to make do with the brain we are born with, the one that evolved with our species. And it

is only by repetition that the brain can be taught a new skill, or to regard the abstract as concrete.

In particular, the way the human brain works presents us with just one way to achieve sufficient familiarity with numbers to be numerate, that is, quantitatively literate: practice basic arithmetic, at least up to decimals and fractions, until you become proficient.

Just how far into arithmetic you have to go is not at present clear, and may vary from person to person. Thus, educators who agree with my thesis so far might disagree over the necessity of teaching particular techniques, such as pencil-and-paper long division. In purely neurophysiological terms, more is definitely better. The constraining factors are time and sustaining enough motivation on the part of the learner. For there is no escaping the fact that, for most of us, the seemingly endless repetition of a particular task rapidly becomes extremely boring, especially during the initial phases when we seem to be making no progress. (Familiarity can breed not only a sense of concreteness, as I observed a moment ago, but also contempt.)

The only real alternative I know to succumbing to boredom and giving up is to keep your eventual goal uppermost in your mind. One way to do this is to keep reminding yourself of what I call the "wow factor." For all that repetition may be boring, it is really a remarkable feature of the human brain that it can acquire such a wide range of new skills. As we have seen, humans are not alone in being born with some innate mathematical skills—a

math instinct. Some species seem able to acquire some additional new skills through a process of repetitive training. But even for the dogs and cats who live with us and for evolutionarily close relatives such as chimpanzees, the range of those new skills is limited, and the training period is generally far greater than it is for humans. We humans are born with what seems to be a truly unique ability to acquire a virtually endless range of new skills. Surely you owe it to yourself to use that precious gift to your advantage whenever possible.

FURTHER READING

For further reading about mathematics in general, at roughly the same level of treatment as in this book, there are my own two earlier books *Life by the Numbers,* published by John Wiley in 1998, the official companion to the six-part PBS television series of the same name, for which I was an advisor, and *Mathematics: The Science of Patterns: The Search for Order in Life, Mind, and the Universe,* published by W. H. Freeman in 1994 in the Scientific American Library series. Another overview treatment that is well worth getting hold of is Ian Stewart's excellent little volume *Nature's Numbers,* published by Basic Books in 1995.

There are many books on the market that describe the mental capacities of animals. Among the ones on my own bookshelf are *Wild Minds: What Animals Really Think* by Marc Hauser, published by Henry Holt in 2000; *Animal Minds: Beyond Cognition to Consciousness* by Donald Griffin, published by the University of Chicago in 1992 (revised edition 2001); and *Apes, Language, and the Human Mind* by Sue Savage-Rumbaugh, Stuart G. Shanker, and Talbot J. Taylor, published by Oxford University Press in 1998.

For excellent coverage of much of the recent work on how the human mind learns and does arithmetic, see *The Number Sense: How the Mind Creates Mathematics* by Stanislas Dehaene, first published by Oxford University Press in 1997, and *The Mathematical Brain* by Brian Butterworth, first published in the UK by Macmillan in 1999 and subsequently published in the U.S. by the Free Press with the different title *What Counts: How Every Brain is Hardwired for Math.*

My earlier book *The Math Gene: How Mathematical Thinking Evolved and Why Numbers Are Like Gossip,* referred to several times throughout this book, was first published by Basic Books in 2000.

For a highly readable account of how people tend to perform poorly when dealing with numbers, see John Allen

Paulos's bestselling book *Innumeracy: Mathematical Illiteracy and Its Consequences,* first published by Hill and Wang in 1988.

For further reading on some of the material in chapter 6, see Mario Levy's book *The Golden Ratio: The Story of Phi, the Extraordinary Number of Nature, Art, and Beauty,* published by Review in 2002.

My principal source for the material of vision in chapter 8 was Steven Pinker's book *How the Mind Works,* first published by W. W. Norton & Co. in 1997.

I drew much of my material on street mathematics (chapter 10) from the book *Street Mathematics and School Mathematics* by Terezinha Nunes, Analucia Dias Schliemann, and David William Carraher, published by Cambridge University Press in 1993. This book is aimed more at schoolteachers rather than a general audience.

The book I refer to in chapter 13, *Where Mathematics Comes From: How the Embodied Mind Brings Mathematics into Being* by George Lakoff and Rafael Núñez, which can be viewed as a sequel to *The Math Gene* (although it was not written as such) was first published by Basic Books in 2000.

INDEX